AF796910

FROM VIETNAM TO OPERATION WELCOME HOME

LT. COL. MIKE JACKSON
&
KAREN JACKSON

Every page of this book is dedicated to the brave men and women who served in Vietnam and came home to lead productive and meaningful lives. Knowing them back then enriched my life. Meeting them again thirty years later renewed my soul.

CONTENTS

Preface ---------- 1

Chapter 1 Looking Back ---------- 4

Chapter 2 Coming Home ---------- 14

Chapter 3 From Two Engines to Eight ---------- 20

Chapter 4 From Career Gloom to Career Resume ---------- 45

Chapter 5 Top Rung of the Ladder ---------- 76

Chapter 6 First One School, Then Another ---------- 94

Chapter 7 Raise Your Right Hand and Repeat After Me ---------- 108

Chapter 8 Back to School Once More ---------- 127

Chapter 9 Tiny Office to Robust National Organization ---------- 142

Chapter 10 The Book That Inspired the Celebration ---------- 168

Chapter 11 The First National Operation Welcome Home ---------- 176

Chapter 12 The Ripples Widen ---------- 189

Chapter 13 Indiana and Beyond ---------- 199

Chapter 14 Recognition for Veterans Grows ---------- 207

Chapter 15 Final Thoughts ---------- 214

Glossary ---------- 219

Acknowledgments ---------- 222

About the Authors ---------- 224

Mike Jackson ---------- 224

Karen Jackson ---------- 226

Preface

In 2004, Tara Dixon Engel and I published our first book, *Naked in Da Nang*. It was a love letter of sorts…the story of a guy from the small town of Tipp City, Ohio, who loved the military, the Air Force, his country, his family, and the much-maligned group known as Vietnam veterans.

Naked in Da Nang wasn't simply the story of my adventures in combat. It was the story of a journey taken by everyone who came of age and joined the military in the 60s and early 70s. It was the story of everyone who has ever left home to travel somewhere strange and scary. It was a story *not* of innocence lost but of innocence doggedly clung to, even in the worst of times.

The book was about going to war and discovering that people everywhere are much the same—a mixture of good and bad, and often just funny—and that situations are truly what you make of them. Yes, even combat. My experiences—and certainly my aversion to being shot at—were pretty typical, as I have learned from the veterans with whom I've talked at Operation Welcome Home celebrations and various book-related activities.

Naked in Da Nang was a tribute to some of the finest people I've ever known and to an era that shaped me and defined the rest of my life in a positive way. It started out as a book about funny things that happened to me in Vietnam, but it became something much more. It became a catalyst that allowed me to do things I never dreamed possible and to mend some things that had been broken for myself and for many others.

As we wrote *Naked in Da Nang*, we interviewed many of the people I served with in the Air Force, including those with whom I had served in Vietnam. After interviewing each of these, I'd shake their hand, and we'd wish each other a sincere "Welcome home." Tara was confused by this interaction since neither the interviewee nor I had gone anywhere. It was a difficult, emotional explanation, and the emotion I felt as I tried to explain my feelings is one shared by every Vietnam veteran I have ever met. "We were never welcomed home," I began, "so we welcome each other." Then she understood.

Over the months following publication, Tara and I often talked about the lack of a proper welcome at the time Vietnam veterans returned home. As time passed and we shared our thoughts, we began to wonder if it was not too late to try to correct, at least in small part, the injustice these brave men and women had felt for more than thirty years. *It's never too late to say "Thank you,"* we agreed.

Thanks in large part to the success of *Naked in Da Nang*, Tara and I did figure out how to welcome home thousands of Vietnam veterans and, later, veterans of all eras.

But as we wrapped up *Naked in Da Nang* and later completed our involvement in Operation Welcome Home celebrations, I knew I still had many more stories to tell—stories from the time when I returned from Vietnam until my retirement from the Air Force. Only a very small percentage of Americans serve in the military today, so fewer and fewer people know what it's like. I wanted to tell the *whole* story of what serving in the military was like… at least for me: the jobs I did, the people I met, the locations I got to know, and the life-long bonds I made with people who share my experiences and values. At the same time, one cannot ignore the hardships endured by your family during a twenty-three-year career.

I spent twenty-three years of my life with exceptional, wild, and wonderful folks, and the wealth of experiences and opportunities energized my career and its aftermath. But my stories didn't end with my Air Force retirement, either. I wanted to tell even more, especially about the amazing people I met through my position with the National Aviation Hall of Fame and the myriad of activities on behalf of veterans that included the Operation Welcome Home celebrations.

The perfect co-writer for these stories is my wife, Karen, a retired language arts teacher. No one else followed me around the country as I moved to every new assignment, whether she was ready to move or not. No one else knew the vast amount of effort that took. And no one else kept our home life stable while I focused on my Air Force duties.

What follows is the story of how I went from being literally naked in Da Nang (a true, clean story) to Operation Welcome Home thirty years later. I hope you enjoy reading about that journey as much as I enjoyed living it.

Chapter 1

Looking Back

As I was growing up, I don't remember a time when I didn't dream of being an Air Force pilot.

I built model airplanes instead of cars like most of my friends did. I practiced operating the "flight controls" using the knobs on my neighbor's kitchen stove. (That stove was SO cool!) On some summer days, I even convinced my father to let me ride with him to Wright-Patterson AFB when he went to work, and I'd spend the day exploring the Air Force Museum until he could pick me up at the end of his work day.

But for years, I didn't let myself seriously consider it because I just didn't think I was smart enough. There are many folks in my past, and some even today, who might agree with me. However, as I graduated from college and the war in Vietnam showed no sign of ending soon, I was smart enough to consider my options. *I can wait to be drafted into the Army and possibly find myself in the jungles of Southeast Asia, or I can chart my own course. What should I do?*

My love for the military, and the Air Force, in particular, was gifted to me by my father, Ed Jackson. Dad had seen some horrific events while serving as a waist gunner in B-24s during World War II. But he managed to keep a positive outlook on life, and he made sure his eldest son shared that outlook. Fortunately, I also had a keen sense of humor, a gift from my mother, who had grown up in the stark poverty of a coal town in southern Kentucky. She was, quite literally, the embodiment of the girl Loretta Lynn sang about in *Coalminer's Daughter*. Despite that, she knew how to laugh, and she shared that knowledge with me. Indeed, I discovered at a very young age that making Mom laugh was the quickest route to forgiveness for my frequent childhood indiscretions. I've attempted to utilize that technique on many superiors through my adult years, but Mom was the only one on whom it worked consistently.

Together, my parents left an indelible mark on my psyche, a mark that has magically shielded me from permanent damage—mental, at least—and gave me the ability to practice gratitude and appreciate the small stuff. Now in my seventh decade of life, I still look forward to simple things like regularly meeting my friends at a Bob Evans restaurant for breakfast. I've never lost that childish sense of anticipation and glee, and somehow, I attribute that to the unique combination of genes from Ed and Winnie Jackson.

As a child, I was intrigued by the pieces of metal shrapnel that danced up and down when Dad flexed and relaxed his forearm. "Big Ed," as we siblings lovingly called him behind his back, dropped out of high school in 1943 to join the Army Air Corps and trained as a gunner on the hulking B-24 Liberator bomber. His adventures included being wounded during the raid on the marshalling yards in Nimes, France, for which he received a Purple Heart. After he was released from the hospital and approved for flight duty again, with his arm still bandaged from that injury, his B-24 was once more assailed by Germans, and he was wounded yet again—Purple Heart number 2. But this time, his plane was shot down, and he and his crew were forced to bail out over Yugoslavia.

Half of the crew was quickly captured, but Dad was one of three who managed to escape and evade, assisted by Yugoslavian partisans who sheltered them for 33 days until an Army pathfinder parachuted into their camp and arranged for them to be picked up by a C-47. You can't make this stuff up. The C-47 landed in the middle of enemy territory on a makeshift grass strip, lit by small piles of burning grass and hay, while P-51s flew cover overhead. Quite a story. Quite a man.

While Dad was convalescing in a Louisville, Kentucky, hospital, he met a beautiful young lady named Winnie Hamblin, whom he married after the ultimate wartime courtship of six weeks. Following an assignment in Liberal, Kansas, Mom and Dad moved to Dad's hometown, Tipp City, Ohio, where Dad finished high school at night while holding full- and part-time jobs during both the work week and on weekends.

As their family grew, I became the oldest of four boys and one girl. While we didn't have many materialistic things growing up, we never knew it. I tell people, "I grew up at the best time, in the best place, and to the best parents ever." I mean it. Tipp City was a small, friendly town where everyone watched out for everyone else. My family lived one block from Tipp City Park, and our only rules were "Be good" and "Be home by dark." That gave me a lot to work with.

I had a wonderful childhood, and the friends I made in school are still some of my best friends. In fact, you can find quite a few of us enjoying breakfast and solving the world's problems in Tipp City restaurants on a regular basis, having seen and done all the far-flung stuff that seemed so enticing and important in our youth—but discovering, in the end, that we had all the stuff that really mattered all along…friends, family, and a little slice of shelter from the storms of life.

After graduating from Tippecanoe High School in 1964, I had absolutely no idea what I wanted to do with the rest of my life, so I followed the crowd and went to college. Okay, so the fact that college was accompanied by a draft deferment wasn't lost on me, either. The conflict in Vietnam was intensifying, and being drafted had become a real possibility. *After four years in college, the conflict will be over, and I'll be safe. After all, the war can't last four more years, right?* Boy, was I naïve!

I opted to enroll at Ohio University in Athens. Some called it "Harvard on the Hocking." This decision turned out to be another of those experiences that shaped my charmed life. I hadn't done a lot of research into colleges. Heck, I wasn't even sure I could get accepted into a college, but I loved OU from the moment I saw it. Nestled in the foothills of the Appalachians, it is a picturesque campus that offered the same sense of security and belonging that I'd left behind in Tipp City—but with new and refreshing independence, as well.

OU would provide me with an education, the skills of a teacher (which I have found unconventional ways to use over the years), and the love of a pretty coed. Karen Shorts was quiet and studious and, at first, wanted little to do with the smart-aleck kid who manned the milk machine in Jefferson Hall's cafeteria. When she saw me flirting with every girl who came through the line, she ignored me and refused, despite my pleas, to even say *Hi*. I was intrigued.

The harder I tried to strike up a conversation with Karen, the more determined she became to ignore me. She tried to go through the other meal line, but I noticed. I let her walk the full length of the cafeteria before alerting my buddy, who checked IDs at that line's entrance to the food. "Hey, Bob! Her meal ticket is for the even-numbered line." *My line*. Bob knew me well enough to know I had a reason for snitching.

Since eating in the only dormitory and via the only line where her meal ticket was valid—plus, she was probably hungry—Karen came back to the correct line, but (*Watch this!) I* shut off the milk machine as she neared and stared at her, still waiting for a simple *Hi*. As she turned and walked away toward her friends, I *might* have heard a very quiet "Jerk!"

For the next couple of weeks, every evening at dinnertime, Karen walked past the milk machine in an attempt to avoid me completely. But that left her only beverage choices coffee, iced tea, or water, and she really wanted milk. So, next, she sent her friends to get a glass for her, but I noticed that, too, of course. *No way will I relent*. As with so many things in my life, I simply kept chipping away at her indifference until she finally surrendered out of exhaustion and came to the milk machine. *Now's my chance!*

"You want to go to the MIA with me on Friday night?" (OU's 25-cent *Movie in the Auditorium*) I was a big spender.

"If I say 'yes,' can I have some milk?"

"Sure."

Pause—long pause—while she considered if it was worth it. If *I* was worth it.

"Okay, I'll go."

Three years later, we were married, and we remain married after more than fifty years.

By the middle of my senior year, as I neared graduation in June of 1968, I knew the Draft Board would soon begin to plan my future for me if I didn't do something for myself, so I examined my options. My first thought: *I could teach high school*, seeing as how I would be graduating with a Bachelor of Science degree in Secondary Education. *Seems like the logical thing to do. That'll extend my juicy deferment, too. Hmmm...*

While I was home on spring break, I spoke to a couple of school administrators about my prospects for getting a teaching position at my alma mater, Tippecanoe High School. They offered me a job on the spot. *Wait. Don't these guys have any recollection of the antics my buddy Joe Catey and I pulled in high school?* "Sounds great!" I told them, but I soon realized that was not what I wanted to do. Not at this point in my life.

I owe much of that decision to a brief stint of student teaching history during the fall semester of my senior year at OU. My mentor was a great guy and a fine teacher, who proudly announced when I met him, "I've spent every single day of my thirty-year career in the same classroom, in the same brick building, in my hometown."

And staring out the same window at Nelsonville, Ohio? The man couldn't have been happier with his lot in life, but I tried to picture myself in his position thirty years from then. Not a picture that appealed to me. I knew I had to take my chances with the whims of fate.

Now, don't get me wrong. The older I get, the more I value consistency and predictability in life. But, at 21 years of age, I was pretty shaken by the prospect of waking up every day for the next thirty years and heading off to the same small classroom in the same brick building, staring out the same window at *my* hometown. Furthermore, the thought of teaching about places I'd never seen in person and experiences I'd never had frightened me. I turned down the teaching contract in June.

Next stop: the Air Force and Navy Recruiting Offices. If I was going into the military, it would be on my own terms. I applied for pilot training with both branches. Being the honest, trustworthy man whom my parents raised, I then visited the Draft Board in Troy, Ohio, and informed them that I'd decided not to teach school. What a dumb move! I obviously should have waited until I'd received responses to my applications. There went the deferment. "Then, you're 1-A," I was told. Soon after, I received a notice in the mail to report for an induction physical.

I raced quickly back to the Draft Board office to plead my case. "Why send me into the Army for two years when I'm ready and willing to serve six years in the Air Force or Navy? Just wait until I hear from them," I argued. No sympathy; they had quotas to meet, and they were going with a sure thing.

I felt like a deer looking into the barrel of a hunting rifle. On my way out, I slowed to talk to the young secretary, and my weasel personality took control. (After all, my honesty had certainly done me no good.) After we shared some personal information...*Where did you go to school? When did you graduate?* I asked her out. Yes, I was steadily dating the girl I hoped to marry, but she was back at OU, and I was desperate. *There's a perfectly good reason for these tactics,* I convinced myself. *Karen will never find out, anyway. Right?*

So, after a date or two, the young, understanding secretary pulled my file until after the papers were processed for me to enter the Air Force. Then she put the file back. Yes, it was a sneaky thing to do, and Karen was NOT pleased when she heard my dear buddy Joe joke about the episode years later in the presence of my then-wife. But she had to agree that it saved me from being a grunt when I went to Vietnam. That prospect frightened her more than her displeasure at my antics.

By fall, I was at Officer Training School on Lackland AFB in San Antonio, Texas, and subsequently in pilot training at Laughlin AFB, also in Texas. In December of 1969, I flew to Cleveland one weekend to marry Karen on a Saturday, and we flew to Del Rio on Sunday morning in time for an early take-off Monday morning. Your Air Force never sleeps. We joked that our flight from Cleveland to Del Rio, Texas, was our honeymoon.

Despite this abrupt start, Karen quickly adjusted to the challenges of military life—and there were many—with the grace and dignity that exemplifies the unsung heroism of military spouses. Her quiet strength would become the perfect complement to my manic energy. She put up with it all—my combat tour in Vietnam, the frequent relocations across the country, leaving good friends and familiar surroundings behind, settling into new housing, and seeking new teaching positions for herself. Then, just as we seemed to get all the pictures hung, we'd be off to another Air Force assignment. There's a reason my friends call her "Saint Karen."

I arrived in Vietnam by a somewhat unusual route. When I graduated from pilot training in February of 1970, I headed to Ft. Walton Beach, Florida, for three months, learning to fly the O-2A aircraft at Hurlburt Field. From there, I, along with about a dozen other new pilots, was selected to become fighter qualified, which meant three additional months of training in the AT-33 at Cannon AFB near Clovis, New Mexico. Finally, I reported in September to one of two brand-new stateside forward air control (FAC) squadrons, one at Shaw AFB in South Carolina and the other at Bergstrom AFB in Texas. I ended up at Shaw flying O-2As.

The idea the Air Force brass had come up with was to train some new pilots as forward air controllers (FACs) in a stateside squadron alongside experienced pilots who had recently served one-year tours as FACs in Vietnam. The plan was the newly minted FACs would eventually be sent to either Korea or Vietnam, already experienced as forward air controllers and, thus, combat-ready. We would arrive in-country with the skills and experience to hit the ground running without requiring the typical month or more of check-out flights with an instructor pilot. We would be, the Air Force reasoned, fully prepared for combat. I suppose it made sense, but it's tough to prepare anyone for combat without actually placing them in combat.

Then, much too soon, we were told we were ready, and, much to Karen's chagrin, I volunteered for Vietnam. After a short vacation trip with Karen through New England and a move for her to her parents' home in Cleveland for a year, I set forth.

I arrived in Vietnam after the requisite two weeks of jungle survival training in the Philippines, known affectionately as "snake school," with my hand stitched up and in a splint thanks to a sharp piece of bamboo unleashed by an overly enthusiastic classmate's machete. Arriving in a combat zone already injured is not a good omen.

Following a couple of days of in-processing at the USAF base at Cam Ranh Bay, Republic of Vietnam, I was sent to Da Nang Air Base in northern South Vietnam to become a part of the 20th Tactical Air Support Squadron. Eventually, I headed to Camp Eagle, the headquarters for the Army's101st Airborne Division. Yes, I said ARMY. OK, so let's look at this situation. I'm an Air Force pilot flying a glorified civilian aircraft with no armor or ejection seat...but at least I'm about as far north in the war zone as anyone can be, and somehow, I managed to be one of fewer than ten Air Force people living with thousands of Army troops. It just kept getting better and better. *How did this happen?* I pondered. My father had lectured me, "Join the Air Force. You'll have hot meals and clean sheets to come back to at night." I listened. I joined the Air Force. And here I was, living in the dust and heat with a bunch of Army guys who seemed to like to get dirty and sweat profusely. Lucky me.

But, what the heck, I was nonetheless eager to get to work and start earning my flight and combat pay. And, earn it, we did! Our little group of forward air controllers was responsible for controlling all Air Force, Navy, and Marine airstrikes from the demilitarized zone (DMZ) between North and South Vietnam south to the outskirts of Da Nang and from Laos on the west to the South China Sea on the east. It was an extremely active and dangerous area of the war zone.

There's just no substitute for the real thing, so while I may have been as informed, trained, and educated on the rigors of combat as possible, I was definitely *not* prepared mentally. Indeed, *Naked in Da Nang* got its title not just from an unfortunate showering incident during a typhoon but also from the exposure we all felt. We were naked to this new experience called war, unprepared for the threats and challenges that arise when you are plopped into a strange country where people want to kill you.

That year was a challenge for both Karen and me, as well as for my parents and siblings. Communication was nothing like it is today – no cell phones, no email, no Facebook, no

Skyping…simply no instant access of any kind. We wrote letters, occasionally inserting a cassette tape. My dad tracked my locations on a giant map of Southeast Asia which he hung on the basement wall. It was probably his way of maintaining some kind of control, but my sister told me years later, "I hated that map!" because it seemed to embody all the uncertainty and angst of having a big brother in a war zone.

In some ways, it was easier for me than for them. I KNEW I was still alive, reasonably healthy, and intact. However, Karen and my parents received only periodic (and wildly inconsistent) delivery of my letters to assure them of my safety. Don't get me wrong; I wrote regularly. In fact, it was the letters that Karen saved from my year in combat that years later helped me reconstruct much of the timeline for *Naked In Da Nang*. The problem was that the military wasn't in the business of mail delivery; they were in the business of waging war. Thus, my heartfelt writing could travel to some undisclosed location for days, or sometimes even weeks on end, only to show up on my family's doorsteps in a bundle…just as my loved ones were certain an Air Force car with a chaplain would pull up to their curb at any moment.

My year in Vietnam lasted from June 25, 1971, to June 25, 1972 – and included a leap year with the associated extra day. (I had already decided that I was going to be royally ticked off if I died on the 366th day.) I'm not sure that any combat tour of duty is ever routine, and mine was no exception, but I was one of the lucky ones. After flying 210 combat missions, I made it out alive and reasonably intact. However, Vietnam has stayed with me in the form of a combat injury, including severe neck and back issues that still linger. But I'd figured it out. I'd survived.

An O-2A taxiing from a revetment at DaNang

Chapter 2

Coming Home

The engines revved, we accelerated down the runway, gear up… and we were in the air! As we lifted off, thunderous cheers erupted from the passenger cabin. We were heading home.

My departure from Vietnam was via a civilian contract "freedom bird"—a Pan Am 707 that flew us from Tan Son Nhut to Honolulu for a brief refueling stop and, finally, to Travis AFB near San Francisco.

I was seated next to my boss in Vietnam, Major Dale Ullrich. We had served together from nearly my first days at Camp Eagle until my final days at Da Nang. Dale became more than just my boss; he was my friend. *I'll miss him*, I thought, as I looked past him and out the window at the lush, green Vietnamese countryside rushing by. We'd been through a lot together.

"The first thing I'm gonna do when I get home is have a beer and sit in front of an air conditioner. Just sit there," Dale said. "Then I'm gonna take an hour-long shower. Next, I'm gonna put on something that isn't olive drab or camouflaged. How about you, Jackson?"

I grinned and raised an eyebrow suggestively.

"Yeesh! You newlyweds." We laughed.

I closed my eyes and visualized the faces of my wife and parents, who'd be meeting me at Dayton International tomorrow. Boy, it's going to be good to see them again!

But I also thought about the troops we'd left behind. None of us were macho warriors—just a great group of normal guys. We'd been sent to do a job, and we'd done it to the best of our ability.

But mostly, I slept on that leg.

As we felt the thud of landing gear on asphalt in Honolulu, cheers filled the cabin again. As the thrust reversers whined and the plane began to slow, I viewed the Hawaiian landscape rolling past my window. I felt almost giddy. *Yes! I'm back in my own country, back among my own people.* A huge smile crossed my face. *I'm back.*

Once on board our continuing flight to Travis AFB, the talk included what we might find when we reached San Francisco. We'd seen the growing anti-war sentiments reflected in the media before we left for Southeast Asia a year earlier. We hadn't been living in a vacuum for the last year, either. We knew the country was seriously divided over the war.

Yes, the atmosphere upon our upcoming landing in California was unknown to us. As our fellow passengers shared what they'd heard about the homecoming of earlier troops, Major Ullrich tried to put things in perspective for me. "Don't know what we can expect. Not even sure I care at this point. At least we're home, Jackson. Remember that."

The flight from Hawaii was shorter than the leg out of Da Nang, so it didn't seem long before our aircraft began its final descent into the Bay area and landed at Travis. Then the loudest cheers of all broke out. We couldn't have been happier to be on U.S. soil again.

You did it, Jackson! You actually made it out alive! You're going to go to bed tomorrow night in your own bed, in your own town, in your own country. You won't have to listen for the whistle of incoming mortar rounds and rockets. No more wondering if your next meal will be your last. You're one of the lucky ones... you're actually going to get your life back.

As we eagerly deboarded, we assembled in the base operations terminal to be advised on ground transportation. "You can take an Air Force bus to San Francisco International or use other modes of travel to your next destination."

I didn't think anything could dampen our spirits. Boy, was I wrong! The briefing item that caught all of our attention was being told, "If you're going to fly out of San Francisco International, you might consider first taking off your uniform and putting on civilian clothes."

Actually, this wasn't an unexpected warning, as we'd all heard about the unwashed, rabid, anti-war protesters who hovered outside and inside "the dreaded SFO terminal" to welcome home the GIs returning from Vietnam and to harass those troops on their way over. We knew about the protestors, but it stung and pissed us off royally to actually hear it from an Air Force briefer. That still pisses me off.

Dale and I would be taking different buses based on our final destinations, so we said our farewells at Travis. The goodbyes were tinged with sadness, despite our mutual relief at heading off on the final legs of our journeys home, Dale to Texas and me to Ohio. You don't live with someone for the better part of a year—surviving daily attempts on your lives—without developing a bond that can't be broken. Fortunately for me, Dale was also continuing his Air Force career. I felt sure our paths would cross again.

For me and for all of us returning from Vietnam, we felt we had done our duty and risked our lives for our country, and any anti-everything filthy protesters weren't going to intimidate us. I don't recall any of us changing clothes, and we ALL took the military bus from Travis AFB to SFO in our sweaty, two-day-old uniforms.

I caught a glimpse of the San Francisco airport's control tower in the distance and felt my stomach jump… in a good way this time. I watched a commercial airliner disappear against the horizon and grinned, realizing I'd be on one just like it in a few short hours. It would take a lot to spoil my optimism today.

Then, seemingly from out of nowhere, they materialized. It was a rag-tag group of five or six waiting for us, and the chanting, waving of signs, spitting, and attempts at intimidation began as soon as the bus pulled up to the curb. I've heard it said by some in the years since my return that this sort of "welcome" didn't really happen. But, yes, it did. I was there. I think most of us returning veterans would never understand why these losers were harassing fellow Americans who had no voice in any war policy when they should have been staking out the airports in Washington, D.C., to protest the politicians who voted for those policies.

Those protestors have shaped my negative opinion toward most demonstrators—and liberals in general—for the rest of my life. But, on a positive note, the protestors did help unite Vietnam veterans into a closed group who felt they were brothers and sisters who might not trust the system but could trust each other completely.

Yes, it was an interesting era, to put it generously, but I can honestly say that the positive memories from my tour of duty outweigh the negative ones. The guys I served with were honorable, courageous men committed to doing the best they could under less-than-optimum circumstances.

All I wanted to do was get out of the terminal lobby, board a flight heading east, and fly home to Karen and my family on Tippecanoe Drive in good old Tipp City. I was in the United States. No one was shooting at me.

Forget the hippies. You've got two more flights. Just a few hours until you can start living your life again. It's done. You're going home. Life is good. Really good.

And things definitely improved. Karen, Mom, and Dad met me at the airport, and I arrived in Tipp City to see the hand-drawn signs my brother Don and sister Debbie had placed in the yard depicting Snoopy, my favorite cartoon character, welcoming me home.

The Vietnam assignment was over, and I was ready to continue my life in peace and quiet. But, while I was home from Vietnam, I came to realize that Vietnam was always going to be a part of me, and so were my fellow Vietnam veterans.

Then-Major Dale Ullrich, one of my FAC supervisors, and I flew back from Vietnam together on our "freedom bird." Colonel Ullrich was to help guide my USAF career for many years to come.

Welcome home signs in my parents' front yard greeted me when I arrived home.

Chapter 3

From Two Engines to Eight

So, I was heading home from Vietnam and about to enter the next phase of my Air Force career, which I was fairly certain would be the *last* phase when I found out my next assignment. I'm not normally one to complain. Really, I'm not.

I'd been assigned to Tactical Air Command (TAC) in my 0-2 assignment at Shaw AFB, and I was verbally promised a TAC fighter as an incentive to volunteer to go fight the war in Vietnam rather than being sent to Korea. For the entire year I spent in Vietnam, I was looking forward to flying a fighter jet upon my return home.

But that wouldn't happen now. When the assignment came down, it was a B-52. That verbal promise meant nothing. I was promised a Porsche but delivered a semi-truck!

I should have seen it coming. My buddies from the 704th TASS at Shaw—Tom O'Neill, Lyndle Price, Ted Kolias, and Stan Buczak—who served in Vietnam at the same time I did were getting either pilot-training aircraft in Air Training Command (ATC) or SAC's B-52s and KC-135s (air refueling tankers), so it really didn't surprise me to get a B-52. But it didn't make me happy. In fact, I was outraged.

I truly understood that, like everyone else, I served according to the needs of the Air Force, but I felt as if I'd been lied to by the service I so loved. *This is a perfect example of bait and switch!* I thought. *I did everything the Air Force asked me to do, including risking my life for a year. How can they do this to me?*

At that moment, my love affair with the Air Force moved well beyond the honeymoon phase. I was plenty disillusioned.

Aside from my frustration with "promises made and promises NOT kept," I wasn't at all comfortable at the thought of going into this aircraft. The B-52 is large and complex. It is a HUGE aircraft that weighs up to 488,000 pounds, has eight engines—ten if you count the two missiles that are mounted on a regular basis—and it carries a crew of six. It has as many engines as all of the planes I had flown before put together! That's eight engines on one big, heavy aircraft. Plus, I was used to flying by myself. Could I really even do this?

And I knew going into SAC on an aircrew meant pulling alert, which basically consisted of living underground and being trapped within the alert facility for one week at a time while waiting for a nuclear attack.

Maybe there is something I can do to avoid this, I thought. I felt I had to try, so, in a brazen move—a REALLY brazen move—I placed a long-distance call from my unit in Vietnam to the Military Personnel Center (MPC) at Randolph AFB near San Antonio in an attempt to talk my way out of this. (Hey! I'd often succeeded at this before as a civilian.)

"Hello. I'm Captain Mike Jackson, and I'm calling from Vietnam. I've received my follow-on assignment, and I think there's been a mistake. I was supposed to get a fighter when I return to the states, and I've been told I'll be flying a B-52. That's GOT to be an error."

Silence on the other end. Finally, "Give me a minute to check your records, sir… No, that's correct."

"Uh… would it be possible to *change* my assignment?"

"No, sir. Can't do that."

"Is there anything I can say or do to change this?"

"Afraid not, sir."

Rats.

After brooding for a few days, I bravely placed a second call, this time to the personnel office at the Strategic Air Command's headquarters at Offutt AFB in Nebraska. In a stroke of genius, I lied.

"Hello, I'm Captain Mike Jackson, and I'll be leaving Vietnam for home in a few weeks. I just received my next assignment. I am *thrilled* to be assigned a B-52, but I have one *small* request: Could you send me to Wright-Patterson?"

What I *didn't* tell the folks at Offutt was my reason. If I relocated near my hometown, I could research future civilian career opportunities in preparation for leaving the Air Force as soon as possible. Plus, Karen was also from Ohio, and she would be closer to her home, and, equally important, she already had certification to teach in Ohio. That might come in handy while I looked for gainful employment. It would be the best location for both of us, even though it wouldn't be the best location to pull alert.

Having grown up near Wright-Patterson, I was somewhat familiar with many of the operations there. The SAC Wing was isolated from the rest of the base at a sufficient distance that aircrews on alert were not permitted to leave an extremely limited area, lest the klaxon sounded and we had to run to the planes.

Of course, no one could go home during the week while on alert. That was true at every SAC base. But, heck! At WPAFB, you couldn't even take a trip to the base theater, the base exchange, the bowling alley, the hobby shops, etc., to pass the time. Just too far. *Those gates and fences around the SAC facilities are installed to keep the alert crews in, not to keep saboteurs out.* Made sense to me.

At any rate, the personnel officer on the line was stunned. I'm pretty sure *no* former TAC pilot had EVER called to say they were happy to get a B-52 assignment. He should have wondered if I were mentally unstable and should be trusted to fly a plane that could carry nuclear weapons. Instead, I was soon assigned to Wright-Patterson. Cha-ching!

But, being based at Wright-Patterson AFB near Dayton, Ohio, with an assignment flying the B-52H presented me with a quandary. On the one hand, I was obviously thrilled to be back in the United States, and being based near my hometown was great. On the other hand, I did NOT relish the idea of flying a B-52 or serving in the Strategic Air Command (SAC) in any capacity, no matter how long or short the duration. (Sorry, SAC guys.)

I know now that I arrived at Wright-Patterson with a BAD attitude, and that was unlike me. I'd always enjoyed serving in the Air Force and found positives in every job and every location. The fact that I also had a bad back, a bad neck, a bad stomach, and probably a touch of PTSD (before the term was invented)—souvenirs of Vietnam—didn't help, I suppose.

I signed into the 34th Bomb Squadron, 17th Bomb Wing, Strategic Air Command shortly after I arrived back in Ohio in June of 1972. I figured they probably wouldn't have much for me to do if I signed in early since I wasn't trained in anything yet, and it would save me some leave time. Sure enough, all I had to do was show up once a day, say *Hi* to the operations officer, and ask if he had anything for me to do. *Of course not; I know nothing about this job!* Then I'd hang around for a little until a few people saw me; then, I could go home if I wanted to. It gave me time to get settled before leaving for more than three months of B-52 flight training.

While I was putting in my final days in Vietnam, my parents found Karen and me an apartment to rent in Troy, near Tipp City. This wasn't as easy as it might sound since we knew I would be in Ohio for only a few weeks before heading off to Castle AFB in California for three months, plus another couple weeks at Carswell AFB in Texas for nuclear weapons school along the way. Finding an apartment that would allow us to rent for one month at a time wasn't easy, so Karen and I appreciated the help.

I didn't know until years afterwards that Dad was so worried about my being targeted by anti-war protestors (he called them "goons") that he would regularly drive through the parking lot of the apartment complex at any hour of the day or night to make sure no one was vandalizing our car, which had a WPAFB access sticker on the front bumper. As I've mentioned, the military was treated not too kindly during the Vietnam War years.

Our next task was to schedule delivery of the household goods we had placed into storage for thirteen months when we moved from base housing at Shaw AFB. Since Karen would be living with her parents in Cleveland, and I would be living in a hooch in Vietnam, neither of us had much need for the limited number of items we had accumulated during our first year of marriage, but we were eager to access them again. Since having them packed at Shaw was an unpleasant experience from start to finish, we didn't know what we'd find.

Neither Karen nor I had much experience with moving. Everything we owned was packed in a few boxes when we'd left Del Rio, and we hadn't acquired much more since. But, since the Air Force was contracting with professional movers and paying for these services, who were we to question the process?

Packing day in our base housing started off badly and grew progressively worse. I began to worry about the packers' efficiency (and honesty) as two grown men took all day to pack, even though we owned so few things, and they used very little packing paper. And they were SO slow at that.

Frustrated to no end, I left Karen in mid-afternoon to supervise as best she could while I stepped out and drove to the base household goods office to ask them, "PLEASE could you send out an inspector to guarantee things are moving along as they should?"

"We're swamped today," I was told. "This is our busiest time of the year. There's no one available. Sorry."

During the short drive home, I wondered if my being a lieutenant was a factor. I couldn't imagine a Colonel or General getting the same response to their plea for help. Not much I could do to change that, though. I'd just need to stay close and keep my eyes open. Our spare bedroom was my biggest concern, so I hung close. Eventually, everything was boxed and loaded on a truck to be transported to a warehouse until my return from Vietnam.

Now that we had a new residence, be it temporary, I scheduled the shipment from South Carolina to our apartment so we could pull out what we wanted to take with us to California and temporarily leave the remainder in my parents' basement.

As happens to most military people at least once during their careers, our move didn't go well. We checked the inventory as items entered the door and quickly found that four of the larger cartons had "mysteriously disappeared" while they were in storage. Worse yet, the warehouse where our household goods were stored had flooded sometime that year, and many of our possessions, such as a large area rug and several boxes of clothing, arrived either wet or mildewed. My Air Force rain jacket had sprouted green fuzz. And where was Karen's clock radio or any of my sweaters? We wouldn't be driving back from California until mid-December. I might need those.

I tried to gain some perspective before I lost my temper with the innocent delivery truck driver. I kept telling myself, *I'm home, and no one is shooting at me. Plus, it could have been worse. Some moving vans catch fire, and the owners lose all of their possessions.* We owned very little to damage or lose.

The next time the Air Force moved us, if they ever got the chance, we would know better. We would take more control from start to finish. But how can anyone control what happens in a warehouse several states away? *Just save what you can, place a claim for the remainder, and move on.*

The week after I arrived home, it was the Fourth of July, and Karen wanted us to go to Tipp City Park to watch the fireworks. She was sure I'd welcome the chance to celebrate America's independence and possibly run across some hometown friends at the same time. I hemmed and hawed for a while, and I finally declined.

"Why?" Karen queried with eyebrows raised.

"I just don't think I'm ready for firecrackers and loud noises right now."

Karen considered that for a moment, but she understood, of course, so she counted on my going the next year. That wasn't going to happen. Even fifty years later, I rarely go to fireworks displays, although I did go when our daughters were young. Any sudden, sharp noises still take me back after all this time. I would guess that anyone who has ever been in combat would tell you the same thing. Non-military people I know don't understand why I have such a problem with fireworks, but I'm certain many veterans do. At least those who have been in the thick of things.

As I've said, I had quite a bit of free time, and I spent much of it around our apartment. One day, while Karen was making up the bed, I entered the bedroom and noticed the upright vacuum cleaner off to the side. *Hey! Watch this*, I thought. *This will be good for a laugh.*

With Karen's back to me, I plugged in the vacuum, raised the power motor above her head, and turned it on. What a shriek I heard! I'm surprised no one in our apartment complex called 911 to report a possibly-deadly crime.

"Turn it off! Turn it off!" Karen screamed.

Unbeknownst to me, an upright vacuum has motorized rollers with brushes that rotate to help suck up whatever they contact—including long, brown hair. My parents' canister vacuum didn't have rotating brushes. With Karen sobbing, I quickly attempted to untangle the mess I'd made.

"I'm Sorry! I'm sorry!" I repented. "I didn't know there were brushes in it! I thought I'd just make your hair stand on end, not grab it and trap it!" *Yeesh! What an idiot I am!*

Karen failed to see the humor I'd expected her to, and, with both of us working together, we eventually untangled most of her pretty hair, leaving some remnants behind, wrapped around those darn brushes. I'm not sure she's ever forgiven me. (Remember I said my friends call her "Saint Karen?") From that day forward, I think she's wanted me to do all of the vacuuming in not only that apartment but every residence since as penance.

While we were waiting to leave for the B-52 training, Karen and I explored the Tipp City real estate market and found a suitable, affordable ranch-style house in the township that surrounded Tipp City. One good thing about my year in Vietnam—maybe the ONLY good thing—was we saved money. Between Karen living with her parents and banking all of her teacher's salary, and my living in a hooch with nowhere to spend money other than by ordering stereo equipment and a camera from the PAC-EX (Pacific Exchange) catalog, we accumulated enough savings to pay off my college debt and have enough remaining for a down payment on a house. *Thank you, Vietnam.*

This would be our first house following a string of rental apartments and base housing. Exciting! There was one small complication: we didn't own much furniture since we'd been nomads up to then, always dwelling in furnished places. A sofa and two end tables were the extent of it. So, prior to leaving for training, we ordered the basic needs—a bedroom set, a kitchen table, and four dining chairs. Once we'd returned from the three months of training, we figured we could use our webbed aluminum lawn chairs until we could afford real furniture for the rest of the house. It was a reasonable plan, and, to this day, we've never bought furnishings on credit.

In mid-August of 1972, Karen and I packed up our yellow 1970 Volkswagen Beetle and headed off to Carswell Air Force Base outside of Fort Worth, Texas, to attend the Strategic Air Command's nuclear weapons delivery course for B-52 pilots and navigators. That actually turned out to be fun in that we were housed in a nice hotel that the government had contracted for students to stay in, and the course itself, while complicated, was both interesting and not that difficult for me to understand. *Maybe I'll be able to handle this.*

After two weeks of weapons training and plenty of time to lie around the hotel pool, Karen and I packed up our Bug again and headed to California for B-52 aircrew training. That would be a little more demanding, but we were excited, since neither of us had ever been to California before, other than when we passed through an airport on our way to and from Hawaii.

B-52 aircrew training was taught at Castle AFB near Merced, right in the middle of the San Joaquin Valley. "This looks like West Texas!" Karen said as we drew closer and closer.

"Yeah. Sure doesn't look like the California we see on TV and in the movies."

All around us were almond orchards and irrigated farms where much of our nation's food supply was grown. There were tons of pickup trucks, tractors, cowboy hats, and cowboy boots in all directions. Yep, it sure looked like Texas, at least until you drove an hour east to the mountains of Yosemite National Park or northwest to the San Francisco Bay area. It would have been an ideal place to go to school – if it hadn't been B-52 school.

As with the weapons school in Fort Worth, the Air Force provided us with a nice place to stay near Merced. The on-base temporary quarters were full, so many trainees lived in off-base contract duplexes which the government had arranged. As fate would have it, I was assigned the left side of a duplex, and a friend who had served in my squadron in Vietnam, a fellow 0-2 pilot and BILK FAC, Fred Barlow, lived to our right. Many of the guys going through training were Southeast Asia returnees and, to a person, were quite upset to have gotten a B-52 assignment, so we had a lot in common from the start.

The training to fly such a huge aircraft with so many new systems was intense, and Karen spent long hours in the evening quizzing me on emergency procedures, but weekends were ours to explore California. From panning for gold in the rivers of Yosemite to weekends seeing the sites of San Francisco to visiting the home of Mickey and Minnie Mouse at Disneyland on Thanksgiving weekend… We took full advantage of the opportunities.

Since my prior military flying had all been in small aircraft, just riding in a B-52 was a totally new experience for me. I will never forget my first low-level training mission at Castle. The B-52 was designed way back in the late 1940s as a high-altitude bomber which would—in those days—fly so high and so fast, using highly advanced electronic defenses, that nothing could shoot it down. Well, that was a great theory, and it might have been true at that time, but the Russians quickly came up with better missiles and fighters to negate many of the early B-52's advantages.

As a result, the planning gurus in the Department of Defense, working with Air Force engineers and weapons experts, decided the best way to counteract the increased threats when flying at high altitude was to fly to and through the target area at a low level. Great idea, in theory. But, flying an ungainly, 488,000-pound, 160-foot-long, eight-engine bomber at high airspeed at a low altitude didn't totally translate into the nimble machine they envisioned. Hardly.

So, on this occasion, we were flying with the altitude curser set at 500 feet above the ground and holding an airspeed of 325 knots while flying through a 5,500-foot-high mountain range. The ride was extremely bumpy, as you can imagine, and I think that all of us "newbies" on the training flight were wondering just how good this much turbulence was for the old B-52. (Also, how secure the fillings in our teeth were.)

The only crewmembers who can see out of this aircraft in flight are the pilot, copilot, and, on this training mission, our instructor pilot. The two navigators and their instructor are seated below the flight deck in the "basement"—or "wine cellar," as we nicknamed it—and can see only the gauges and radar scopes in front of them. The electronic warfare officer faces backwards in the upper deck and can see only his gauges. In the "E" model B-52 we used in training at Castle, the gunner was wedged into the tail of the aircraft with a great view, but it was out the back of the plane.

While the other trainee was flying the plane, I loosened my seat belt and shoulder harness to glance back at the wings. The B-52 is so long that the pilots can't see the wings from a normal sitting position—and, as it turns out, that's probably a good thing. As I raised up out of my ejection seat and craned my neck backwards to look out the side window, our instructor pilot said, "You don't want to do that." Of course, that made me want to look even more.

Holy cow! I soon understood why the IP didn't want me to see what the wings were doing...*They're flapping up and down like a bird in flight!* Which, I guess, was kind of appropriate.

The B-52's wings were designed to flex in flight, and they actually did this, to a lesser extent, when taxiing on the ground as well. When the fuel tanks in the wings were full, the outrigger gear on the wingtips rested firmly on the ground, helping to support the weight of the wings. Once we'd landed from a mission, the wing tanks were less full than when we took off, and the outrigger gear on the wingtips could be ten feet or more off the ground.

But enough statistics. Back to my first low-level training mission. When I looked back at the wings, I saw they were rising and falling probably twenty feet at the tips. To make matters even scarier, the engines, which are mounted in pairs on nacelles, were moving slightly back and forth sideways on their mountings. From that day forward, I avoided looking back at the wings during a low-level flight. Better to focus on what lies ahead.

I successfully completed the three-month B-52 flight training course and was eager to return to Ohio, settle in our minimally-furnished house, and form a routine. Karen and I had talked about getting a dog once we were living in a place where we knew we'd stay for a while. The way we'd been moving back and forth across the country every month or two didn't accommodate owning any kind of pet. We both had always liked the *Peanuts* comic strip and Snoopy in particular. We extended that fondness to all beagles. We watched the classified ads in the *Dayton Daily News* until we found what we thought was a perfect choice: a full-bred beagle puppy, and we made an appointment to see him.

"Oh, isn't he adorable?" Karen was smitten. "Why wouldn't everyone want one?"

We soon discovered the reason.

We brought *Barney Dog* home in January of 1973, shortly after moving into our house. A bit of advice: Don't bring a puppy home in the depths of winter in Ohio unless you can be tethered to the house. Karen had started substitute teaching so she could become known as the fantastic teacher she was (I may be more than a little prejudiced on that) and be offered a full-time position. I spent long hours at the base. As a result, Barney was often left alone.

We tried leaving him indoors while we went to work. This was before indoor cages came into use. We first left Barney in our garage, but he tried to get indoors by clawing and chewing through the drywall and the door to the house.

Next, while barricaded in the kitchen, Barney chewed two rungs on our brand-new Ethan Allen dining chairs. He also escaped into the family room one day by pushing aside a plywood barrier, and he pulled some books from a bookcase. The one book he dragged into the kitchen and through his water bowl, causing it to stick to the floor, was *How to Train Your Beagle*. We couldn't decide if he was very smart or very dumb, but we decided we'd made a BIG mistake.

So, Karen placed a "Dog for Sale" ad in the *Troy Daily News*. The first day the ad appeared, a woman called and asked, "Does he hunt?" in a *strong* Kentucky-sounding accent, similar to my mom's relatives. Karen explained that Barney's father was a field champion, but Barney was just a house dog. The caller arranged to come see him.

After Karen hung up the phone, she looked at Barney's sad eyes, and she couldn't do it. She hated the thought of Barney riding around in a cage in the bed of a pickup truck—in any kind of weather--to the boondocks so his new owner could play Great White Hunter. *I can't do that to him*, she realized. So, she and Barney climbed in her Volkswagen Beetle, and the two of them rode around the countryside until Karen was certain the potential purchaser had come and gone from our home. Barney Dog would stay.

But I was having bigger problems than a hyperactive puppy. I could tell I was going to have problems with my neck and back on missions that lasted anywhere from six-to-eight hours or even longer. Wearing a five-pound helmet and an oxygen mask—or even just a headset, alone—while bouncing up, down, and sideways in turbulence or during terrain avoidance missions at low altitude was already causing painful neck aches and headaches that lasted days after each flight.

No way could I spend any more time than absolutely necessary flying the B-52. Both my mental and physical health were at risk. However, I'd picked up an additional commitment with the additional training, so I was stuck. I owed the Air Force three more years.

As the war in Vietnam was winding down, the Air Force found itself with a surplus of pilots. Normally, there's a cycle of *too many pilots* followed by another cycle of *too few pilots*, but this situation was more difficult than usual. The Air Force had been ramping up pilot production significantly for years due to the increased demand in the war. Now that pilots were not going to Vietnam in large numbers, there were too many pilots for the available cockpits. As a result, most units were overloaded with pilots, as well as with most other personnel positions.

People in many occupational skill areas were being separated voluntarily ("Hey, who wants out of the Air Force?") or involuntarily ("Sorry, your skills are no longer required.") I hoped for a quick exit, but it was not to be.

Since I was fresh out of training, I was assigned to the right seat. Being a copilot was just fine with me, even though I had more total flying hours than most pilots going into the aircraft commander program. My reasoning was I had never flown an aircraft bigger than a small jet, and I had flown alone for most of my career, never with a crew. I knew nothing about crew management, and I knew even less about SAC procedures and protocols. Being a co-pilot until I could separate from the Air Force was just fine with me.

When many of us Vietnam returnees went into SAC, we weren't really needed, and aircraft commander positions were held out as rewards for good and faithful SAC service. As a result of my being one of the most experienced copilots timewise, I was able to do quite well, fairly rapidly, and advance. Within one year, I became an instructor copilot, and during the only ORI (Operational Readiness Inspection) that I experienced during my entire time in SAC, I received the only *Highly Qualified* rating in the squadron.

In addition, I was given a regular commission in the Air Force less than a year after I arrived at Wright-Patterson. Despite my lack of enthusiasm for flying B-52s, my SAC career appeared to be on the rise – whether I wanted it or not.

At the same time, however, my physical issues were becoming more and more consistent and harder to hide. I had made certain my injuries weren't well-documented while in Vietnam. *They'll get better over time*, I told myself. *I'm only 25 years old.*

The flight surgeons knew that something was wrong, so I just stopped seeing them when I feared it could become a matter of my being grounded. At one point, I was in traction for three weeks at the Wright-Patterson Medical Center part-time, once a day. It didn't do much good, and I knew I had to stop complaining, or I was going to be grounded for sure.

I can't really explain why I didn't just stop flying and avoid much of the pain. I guess that's due to my *Never Give Up* mentality. But, as my physical condition began to directly affect my ability to function as a pilot in the Air Force, I began to plan on a post-Air Force career. Not that I wanted to stop flying or get out of the Air Force, necessarily, but I knew I wouldn't be able to fly these long, often bumpy missions with a heavy helmet on my head and not do serious, permanent damage to my neck.

Then there was the whole unpleasant topic of nuclear alert—to me, the worst aspect of being on a SAC aircrew.

Explaining the reasons for my total frustration with all aspects of alert duty is going to take some time here, so feel free to skip the next couple of pages if you really don't want to hear me ranting and raving. I understand completely.

I knew that I'd "pull alert" sometimes when I began the job, but I didn't expect it to be as often or as confining as it was. Three or four crews each for B-52s and KC-135s went on alert on a Thursday morning and stayed within the confines of the alert shack and wing area until the following Thursday morning, when they were relieved by other alert aircrews. For the whole week, each six-man crew (there were only male flight crew members in the 1970s and before) were basically "joined at the hip" and had to do everything together.

We slept and ate in the alert facility--a two-story, concrete, partially underground bunker—which had a dining room, rec room, TV area, and meeting room on the first floor. Beneath the first floor were underground sleeping areas where most crewmembers shared a room with another crewmember and used a common bathroom/showering area. The entire complex was heavily guarded and surrounded by fences, floodlights, alarms, motion sensors, armed guards, and guard dogs.

Since we didn't have access to other base facilities at Wright-Patterson, wives, girlfriends, children, and any other visitors were able to visit the alert aircrews solely by meeting at a covered picnic area near the alert facility or by meeting in cars in the parking lot. You can picture the absurdity of this during the nine months of cool to freezing weather in Ohio.

Just prior to the SAC wing at WPAFB being deactivated and closed, the patio was enclosed and heated. That was a big improvement. But, as I said, that was shortly before the unit was totally deactivated. Of course.

Before any crewmember was "allowed" to be placed on alert, he had to be certified on the specific mission to be flown by that crew if the klaxon went off and the Emergency War Order (EWO) was initiated. What does that mean? Well, it's complicated and probably still classified, so I'll try to explain it without giving away any national defense secrets.

In a nutshell, every SAC wing had a set of missions that it was tasked to fly in the event our national leadership decided to execute the Emergency War Order—i.e., World War III had started. Each bomber's mission included instructions for how to reach the target or targets within enemy territory, the enemy defenses that would be faced and how to overcome them, the specifics of the target, how to reach it in the case of various malfunctions, how to identify it with various malfunctions, and how to reach a safe recovery area and land the aircraft. That's about all I can say without worrying about violating various laws. Let's just say it was a complicated preparation.

The Wing Commander or Vice Wing Commander certified each crew by having every member of the crew—aircraft commander, co-pilot, radar navigator, navigator, electronic warfare officer, and gunner—brief his (Remember—only male crew members at that time) responsibilities during the mission in detail. After the crew's detailed briefing, the certifying officer would grill the crew on various problems they could face, equipment malfunctions, enemy threats, and so forth until the certifying officer was certain the crew thoroughly understood the mission and was totally prepared to overcome any problems they might encounter in its execution.

Each briefing lasted a tough, nerve-wracking three to four hours, sometimes longer. Only after being certified was a B-52 flight crew authorized to be placed on alert, formally relieving the prior crew.

Since each mission was vital to the successful execution of the EWO and winning the war, no crew could go off-duty until another crew had assumed their alert position. Each B-52 on alert had the specified load of nuclear weapons on their alert aircraft, was fully fueled, had all of the cockpit switches set, and was ready for an emergency start.

If the order to launch the alert force of B-52 and KC-135 aircraft was given, each base had only a few minutes to get all alert aircraft off the ground before, theoretically, that base would be hit with enemy nuclear missiles and destroyed. All alert aircraft were to perform a MITO launch—minimum interval take-off—of one plane taking off every fifteen seconds. The wake turbulence of the aircraft ahead of you made for a rocky take-off!

My hat goes off to the dedicated SAC professionals who did this for years on end. But my limited patience in the best of circumstances didn't allow me to just go with the flow and press on. *Just let me go up and fly!* I kept thinking. It wasn't the fault of the SAC procedures; I knew then and still believe they were absolutely necessary. I was just the wrong person to be in this job.

Alert during normal times would have been bad enough, but these weren't normal times. B-52 aircrews found their services required in two locations—in a way, in three.

The second location for us was MacDill AFB in Tampa, Florida. This was a satellite location for Wright-Patterson's wing. Most SAC bases had their own satellite—a place where a couple of aircraft were maintained in order to spread out the force geographically and also limit the number of aircraft in any one place. No one wanted an attack to disable all aircraft at once. Thus, two B-52 and two KC-135 crews had to travel to MacDill every week for alert duty in the planes located there.

Actually, MacDill had its advantages, not the least of which was nice sunny weather year-round. Also, MacDill's base facilities were much more accessible to us. Closer proximity meant more freedom for us. I made use of the base hobby shop and became quite proficient at ceramics. We still have two of my creations in our home: a pitcher and bowl that sit in an antique dry sink and a hand-painted pilot caricature that is displayed on a wall of shelving. Not exactly a macho hobby, but it helped pass the time. However, we didn't see any family or friends from back home for that week, of course.

In addition, there was an even worse prospect—that third location I mentioned. The continuing war in Vietnam made heavy use of B-52s in bombing missions called "Arc Light." Many aircrews were needed in Guam and Thailand for those missions, and those crews were sent for temporary duty from the squadrons back in the United States. Usually, two crews from each wing spent three months at a time in Southeast Asia. Most individuals did this only once or twice during their four-year tour of duty, unless they volunteered more often, but this was another example of time away from families—with the prospect of getting shot at again.

I hadn't been flying at Wright-Patterson for very long before the wing scheduler noticed I had not been sent for Arc Light. He called me into his office and told me, "I wanted to let you know that you're going to be sent Arc Light within the next few months."

What? He obviously has no understanding of what I've been through, I assumed. *This is my opportunity to educate him.*

"Maybe you don't understand that I just returned from a *full year* in Vietnam," I explained. "I spent three hundred sixty-six days there, and I have *no* desire to go TDY to Guam or Thailand and fly B-52 missions over the very place I just escaped!"

"That doesn't make any difference. You haven't served duty there in SAC."

"Haven't served duty there? Are you kidding me? You can't send me there until everyone else on an aircrew here has spent at least 366 days there."

He could see I wouldn't let this pass, and somehow, I was never scheduled for Arc Light. Score one small victory for Scoop over SAC insanity.

Fewer crews available to fill alert requirements at the home squadrons resulted in the need to pull an increased number of weeks on alert to fill required missions. The end result was frequent back-to-back alerts. In normal times, an aircrew might have two or three weeks off between alert tours. Again, these weren't normal times, so crews might have only one week off alert before going back on alert again, and that week wasn't "off" in any sense of the word. The week between alerts was when the aircrews flew training flights to stay current in the aircraft. Even one such flight could result in three days of duty, as there was a day of mission planning, the actual flying could stretch into part of two days, depending on the length and takeoff/landing times for the mission, and then there was debriefing. So much for downtime or family time.

But spending so much time with the same six guys led to some genuine bonding, which was nice. It was about the time of my third round of alert duty when my crew felt they knew me well enough to give me an ultimatum. "You can no longer drink hot tea when we're on alert," said my aircraft commander. "From now on, your choices are water, soft drinks, or coffee."

This seemed like an odd request. "You're kidding, right?"

"Nope. We're all tired of nagging you to 'Hurry up! Hurry up!' when we have to go someplace while you dunk a tea bag up and down until it's 'just right.' Enough is enough."

"O-ka-a-a y…But I don't like coffee."

"Too bad. Suck it up."

So, I did, since I had to admit that I saw his point. I could grab a cup of coffee from the large urn in much less time than it took to brew a cup of tea. And, I got used to the taste of coffee fairly quickly—to the point where most mornings these days, I'm found downing a whole pot at a Bob Evans restaurant.

The tedium of endless hours on alert often led the crew—and especially me—toward creative solutions to boredom. Some guys played cards or pool. Some guys read books. Me? I watched a soap opera. So much for the last vestiges of my macho fighter pilot persona… but there was a good reason.

Now that we were settled back in Ohio, Karen was teaching full-time, and she watched a fair amount of daytime TV during the summer. Her return to the classroom at the start of a new school year meant enduring a painful withdrawal from her favorite soap opera, "Another World." Remember, VCRs had not yet been invented, or at least not for home use.

Given the long, boring hours we spent in the alert shack, I gallantly offered to watch the show, take notes, and regale my wife with all the gory details when I saw her again.

What began as a selfless spousal gesture quickly became my obsession. It didn't take long for me to understand the lure of a serial. When you're tethered to a building, whether it's a house full of kids or a shack full of restless aircrews, it's easy to get caught up in *watching* the lives of people who seem to be *doing* things, even if those "things" are conniving, killing, philandering, and coveting!

I was mesmerized by the machinations of the wealthy Matthews family and their poor relations. Each day I drowned my boredom in the little Peyton Place of Bay City.

Will Steve and Alice ever find happiness? Will the evil Rachel and her manipulative father succeed in stealing Steve away from his sweet wife? Will Alice go insane? Am I insane? Yikes.

It became a running joke among the crew.

"Where's Jackson?"

"Don't bother him. He's watching his soap opera."

Okay, enough was enough. These guys needed a healthy dose of reality, and I was just the guy to give it to them. I did what any red-blooded, combat-hardened American male would do: I introduced them to my "friends" in Bay City.

That did it. Now we were *all* hooked. Who knew that, in the darkest days of the Cold War, the fate of the free world was scheduled around the trials and tribulations unfolding each day at 2 p.m. on Channel 2. Now *that's* something Jimmy Stewart didn't deal with in the movie *Strategic Air Command*!

The crew was especially enamored of sweet, guileless Alice…she was blonde and beautiful, with a pert little up-turned nose and just a hint of Audrey Hepburn's sophistication. She was the girl every boy wanted to bring home to Mom. And, in the ultimate act of devotion –and boredom—we formally named her "The Sweetheart of Crew E-10."

As I often do, I tend to take things to the next level. It wasn't enough for us to worship Alice from afar, I reasoned. She needed to *know* that a dedicated group of red-blooded American pilots and navigators were poised, at a moment's notice, to fight for Mom, apple pie, and Alice Frame!

So, I dropped her a note.

Yeah, I'm sure she gets a lot of fan mail, but not like this. Not from a group of Air Force combat crewmembers whiling away the endless hours on alert and dreaming of a few stolen moments in Bay City with Alice Frame.

For once, I was probably right. I often wonder how much actress Jacqueline Courtney's publicist laughed when she or he received my request for individually signed photos for every guy on my crew.

And yet, in true Scoop fashion, the skies opened, and manna fell from Heaven. Within a matter of weeks, I was tearing open a hand-addressed package, jammed with a personalized 3x5 portrait photo of Jacqueline Courtney for each crew position – and the extra, added bonus of an 8x10 glossy photo autographed "To Crew E-10." Hey, it was harmless fun, and I'm sure she enjoyed it, too.

The time came when I had to leave our close-knit crew. Despite my lack of motivation, I was selected to upgrade to Aircraft Commander, which meant two months of additional training back at Castle AFB. I have to admit it was a nice summer vacation in California for Karen and me. Once again, we enjoyed lying by the pool and exploring the area on weekends. Then it was back to the grind.

My discontent with being in SAC and flying B-52s did not diminish the longer I remained in SAC. It only grew deeper. Coupled with the physical problems I was now experiencing on a daily basis, it became clear that SAC and I needed to go our separate ways. I knew I would never be happy flying the BUFF, and I knew that, physically, I would eventually be grounded… that I *should* be grounded. So, it appeared that I had no choice but to leave the Air Force.

I have always admired the people in law enforcement, and it also seemed a logical extension of my desire to protect and defend our country. So, I spoke to a representative of the Federal Bureau of Investigation at their office in Cincinnati about a career as a special agent, but I was told, "We're recruiting only those individuals with either an accounting or a law degree."

Given my dismal performance in freshman accounting at Ohio University—I ended up dropping the class in the first few weeks—I decided a law degree would be my best option. Fortunately, the University of Dayton has an excellent law school—if I could get into it.

Karen was teaching third grade at an elementary school in Troy, so we'd have a (reduced) source of income if I separated from the Air Force. Also, the GI Bill would provide some financial help during my time in school. So, my plan was set: apply to separate from the Air Force, begin studying for the LSAT (the law school version of the SATs for undergraduate school), and then apply for UD's law school.

I went to the base personnel office and filled out the paperwork necessary to leave the Air Force at the conclusion of my commitment. I loved the Air Force, but my two-and-a-half years in the B-52 led me to believe there was no way for me to ever get out of Strategic Air Command and, thus, no way around my resulting physical issues.

The die was cast. Even if I had second thoughts about leaving the Air Force, there was an effort to downsize the military post-Vietnam, so I couldn't have pulled my DOS (Date of Separation) application if I wanted to.

So, my future seemed set. Well…maybe not. My boss from my days as a forward air control (FAC) pilot in Vietnam, Major Dale Ullrich—now Lt. Col. Ullrich—was based at the ATC headquarters at Randolph AFB in San Antonio, Texas. Somehow, he found out about my request to separate from the Air Force as it went through MPC, and he called me at the bomb squadron to ask, "What's going on with your AF career? I thought you liked the Air Force and planned to make it a career."

"I *love* the Air Force," I responded, but I explained how I felt betrayed by the fact that I didn't get the fighter I was promised by volunteering to go to Vietnam. I hated flying the B-52, hated alert, and could not physically take the hours of flying with the strain on my neck and back.

"What would it take to get you to stay in the Air Force? You're the type of officer the military needs."

"I'd have to get out of SAC. I don't see a future for me there," and I explained how I'd need to either get a non-flying staff position or an assignment to an aircraft that flew significantly shorter, less teeth-jarring missions. "Oh, yeah, and no alert!"

It turns out it's helpful to have good friends in the right position. "Where do you want to go, and what do you want to fly?"

I had a bachelor's degree in education, and I'd always thought being an instructor pilot would be a great job. So I said, "I'd love to fly T-37s as an instructor."

After discussing locations—and the opportunities to advance in each location—and talking to Karen about her going back to school to earn her Master's degree—I told Colonel Ullrich, "We'd like to go to Reese AFB." Lubbock, Texas, is the home of Texas Tech University, a large campus with a fine College of Education.

The next issue to be addressed was the official policy of no one being allowed to withdraw a DOS once it was approved. My friend, the Colonel, told me, "Don't worry about it. Just go to the Wright-Patt personnel office next week." So, I did. My request to pull the DOS was already approved. I love that guy!

Next question from the Colonel, "When do you want to move?"

"Tomorrow?" I joked. It was January of 1975, but I figured, *what do I have to lose by asking the impossible?*

Then I realistically asked, "Three months?"

"Plan on leaving in March. Better put your house up for sale."

My career in the Air Force was back! My potential career in law school and the FBI was on hold…. forever, as it turned out.

Did I mention how good it is to have friends in the right place at the right time? This would be the start of my practice of NEVER letting the USAF take control of my career again—and, as you'll see, I never did.

To this day, I give the credit for my making a career in the United States Air Force to Dale Ullrich, Colonel retired. Thank you, Sir!

A B-52H on landing roll with its drag chute deployed

Karen at the controls of a B-52E at Castle AFB during wives' day

B-52 aircrew "scrambling" during an alert exercise

A typical SAC alert facility. Sleeping quarters were through the "tubes" and underground.

One of the autographed photos Jacqueline Courtney sent to our crewmembers at WPAFB

Karen, Mike, and the infamous Barney Dog

Chapter 4

From Career Gloom to Career Resume

While Karen stayed in Ohio to complete the school year, I drove alone to Lubbock, Texas, the home of Reese Air Force Base. I was so excited! I felt as if a huge burden had been lifted from my shoulders. I'd escaped SAC! I really did it! I was eager to begin a new and fun chapter in my life and my Air Force career. As apprehensive as I had been going into the bomb squadron, I felt equally optimistic and enthusiastic about becoming an instructor pilot.

The first day's drive was sunny and pleasant. I was as happy as could be. However, my mood dampened on the second day when I heard what was going on in Vietnam. I left Tipp City on April 29, 1975, and arrived in Lubbock on April 30—the day Saigon fell to the North Vietnamese. Throughout the remainder of my drive to Texas, I listened on the radio to news reports of the fall of South Vietnam.

I was preoccupied with thoughts of my year in Vietnam and the tens of thousands of lives that war had affected for as long as I could remember. And, of course, I thought of the thousands and thousands of brave GIs from all branches of service who suffered mentally and physically as a result of serving in the war, and I thought of the 58,220 military heroes who had lost their lives. And, of course, I thought of their families.

Upon signing in at Reese, I saw that all of the TVs and radios in every office on base were tuned to the news, and there was a genuine feeling of depression throughout the base. Since I didn't have a house yet, I signed into the bachelor officers' quarters (BOQ), unloaded my car, turned on the television, and settled in to watch the news with videos of North Vietnamese tanks rolling into the U.S. Embassy in Saigon. Film clips showed panicked South Vietnamese, our former allies, fleeing the country in military helicopters, in fishing boats, or by whatever means they could find to try to escape the enemy who now controlled their country. It was just so sad on so many levels. Would we—would I—ever be free of the Vietnam War? For me, no. Not entirely.

But, other than my sadness over the collapse of the country for which we had sacrificed so much, things were looking up for me, and my future looked bright. The people around me in my BOQ were primarily student pilots in the undergraduate pilot training program (UPT). When they saw me in my uniform, wearing pilot wings, they were eager to talk to me about life as a pilot, and I was happy to share tales of my trials and tribulations as a student pilot five years earlier. I feel that the stories of my struggles getting through the program gave many of them the encouragement they needed at that time. I knew then that I was going to enjoy this job. Everyone seemed so young, eager, and motivated—such a welcome change from my last assignment.

Before I assumed the duties of an instructor pilot, I needed to do two things. First, I had to requalify as a pilot in the T-37 aircraft. Second, I needed to locate living quarters for Karen and me, either on base (government housing) or in town.

First things first. Each flying squadron had a separate unit for their particular aircraft called the PQ section. (The military loves acronyms.) PQ, as in pilot qualification, was the section whose instructors trained you on either the T-37 or T-38 aircraft. Just because you are a rated pilot in the Air Force doesn't mean you can fly an aircraft for which you haven't been qualified. PQ was a small section of only two or three former instructor pilots whose job was to get newly assigned instructors proficient in their respective airplanes and, thus, certified before attending Pilot Instructor Training (PIT) to learn the methods of an instructor pilot.

Since I was a fairly experienced pilot by this point, I thought the PQ process would be relatively easy. I was relatively wrong. Acquiring the necessary level of proficiency in all areas of flying the T-37—contact, instruments, formation, navigation, and procedural knowledge—was a demanding task, requiring more work and study than I had anticipated. But I was having fun and looking forward to my new job, so life was good. And, there was no alert! Yea!!!

My other initial project was creating a list of houses for Karen to look at when she flew into town over a weekend. We decided since the housing market was booming in Lubbock, we'd invest the money from the house we sold in Ohio to buy a house in Lubbock, since we'd be here, theoretically, for four years.

Lubbock is a nice town, but it took a while for us to get used to the place, particularly to the constant wind and occasional dust storms and tornados. The only hills in town are the highway overpasses, and, standing at the edge of a cotton field, one has a largely unobstructed view in all directions.

A realtor chauffeured us around to a few potential homes on my list, and then we stopped for a quick lunch at a McDonald's. As we sat inside, the sky darkened to the west, and our realtor began eying it warily. She, like other residents just a few years prior, had survived a deadly tornado which inflicted a great deal of damage, including leaving a tall downtown office building permanently twisted. We realized the woman wanted to quickly return to the safety of her own home—and possibly a storm cellar—when she asked us, "Do you mind exploring on your own? And then just let me know if there are additional properties you want me to show you." She seemed quite nervous, so, of course, we agreed.

A storm never materialized—at that time—and finding a house we wanted to buy was relatively easy, so we made our choice, and Karen flew back to Ohio. The following week, I put in an offer, it was accepted, and we were home-owning Texans. I liked being a Texan. It's a great state.

Karen finished up the school year at Hook Elementary in Troy and left a job and fellow teachers she loved. By the end of June in 1975, Karen, Barney Dog, and I were settled in our new house in Lubbock. Once I was pilot-qualified in the T-37, we loaded up the car yet again and headed to Randolph AFB in San Antonio for three months of pilot instructor training.

We found a nice apartment in Universal City near Randolph AFB, and Karen and I quickly fell in love with San Antonio. Barney Dog, unfortunately, fell in love with chewing the woodwork in our apartment, but that's another story…one we're still trying to forget. *How old does he have to get to give up his taste for wood*? We both wondered.

My PIT class was comprised of about sixteen pilots who came from a wide range of flying and aircraft experiences. Some of my classmates had served in Vietnam and were flying their second or third type of aircraft since graduating from pilot training. (This was my fourth.) Others were brand new pilots fresh out of pilot training themselves. We also had a mixture of former fighter and large aircraft pilots. As luck would have it, as a mid-grade Captain, I was the ranking student in our flight and, therefore, the "class commander"—a lofty title signifying not much of anything.

During training, we alternately flew in both seats of the T-37, the left (student) seat and the right (instructor) seat. Not only did we have to fly the plane well, but we were now being taught to fly, instruct, and evaluate at the same time. I've never had any trouble talking—just ask Karen and my friends—so I found the process interesting. I enjoyed trying to tell my "students" the correct methods to perform various maneuvers, such as landing, acrobatics, instrument approaches, formation, navigation, and low-level flying, by offering them a number of techniques I had learned over the years.

The three months at Randolph passed quickly and smoothly. I had found my niche. I did well in the instructor program and graduated as the class's Top Graduate. Not bad for a guy who had experienced my own set of problems and frustrations as an undergraduate pilot training student. After patching and repainting numerous beagle chew spots in our apartment, we were on our way back to Lubbock, Reese AFB, and a promising future.

Karen and I were happy to finally be in our own house again. Barney Dog was happy to have a wooden-privacy-fenced backyard to try to dig out of or eat through in his never-ending attempts to escape.

You'd think that after completing the pilot qualification program before going to three months of pilot instructor training, plus being the top graduate in my PIT class, I'd be a fully qualified Air Training Command instructor pilot. You'd think that, wouldn't you? You'd be wrong.

Now that I was back at Reese to stay, I was placed in the Buddy IP Program. That meant I became an instructor pilot with two students of my own to train, but I was carefully watched by an experienced instructor (whom I out-ranked) and the flight commander (whom I also out-ranked). I understand the need for quality control for new instructor pilots, but, for crying out loud, I was a combat veteran with 210 combat missions, I had completed fighter pilot lead-in training, and I had been a B-52 aircraft commander. Would the training and evaluation ever end? The short answer is *No*.

I was assigned as a line instructor pilot in B Flight and was told that I would move up through the squadron hierarchy fairly quickly due to my rank and experience. As a matter of fact, the first two Captains I worked for were both subordinate to me, based simply on our dates of rank. (I was a more senior Captain than they were.) But none of that made any real difference to me since I knew, barring some major screw-up on my part, I'd be getting my own flight soon.

As is usually the case for new instructors, the first two students I was assigned were very good students. One had a lot of flying time as a civilian, and the other had been a navigator with a combat tour in Vietnam in the back seat of the F-4 before being selected to attend UPT. Both students became good pilots, and the former navigator could almost—almost—fly formation as well as I could since the pilot (front seater) in F-4s had let him do a lot of flying from the backseat.

As a side note, the ex-navigator ended his career as a major general. I told you he was good, or maybe it was my excellent instruction… Probably not.

A huge change I noticed immediately after getting on the flight line was there were more opportunities to make and hang around with friends. That might sound strange, but in SAC, you spent so much time pulling alert as a crew you tended not to hang around the squadron or socialize much when not on duty. I'm sure our living in Tipp City, twenty miles from Wright-Patt, might have had something to do with that situation, but the others in the SAC wing were, basically, in the same boat and lived a fair distance from the base, too.

At Reese and at other ATC assignments later, I found that, despite flying two times and occasionally three times a day, we had time between flights to socialize. In addition, we often met our wives at the Officers' Club on week days—and almost always on Friday evenings. The Club hired country western bands, and it was fun to learn to dance like the Texans we'd become. Just due to the nature of the job, ATC seemed to me to promote more socialization among its members, and Karen and I were young enough to take full advantage of the opportunities. To see me now, my friends would find it hard to believe that we even took disco dancing lessons with fellow IPs and their wives. This place was fun!

After a few months as a "line IP," I was moved into the Assistant Flight Commander position in C flight under Captain Don Lamontagne. As with Captain Phil Bentley, my Commander in B flight, Don and I got along great and were friends in and out of the office. I will say this about Don: He was a total professional, and I don't think I ever saw him really get upset, not even when we moved his desk, photos, books, and everything else he owned from his office into the bathroom while he was on vacation for a week.

Of course, we moved all of my things into his area and even replaced my photo with his on our flight's photo roster. Well, maybe he got a *little* upset when he returned to the office, but not much. Many years later, Don retired as a lieutenant general and truly deserved all three of those stars.

Shortly after I came into C flight, Don left Reese for England to fly F-111s, and I moved into the flight commander's position, this time legitimately and without moving anyone's stuff to the restroom.

As a flight commander, my role in the squadron hierarchy expanded, and, in addition to flying with our flight's students, I began to fly with weaker students from other flights, some of whom were in danger of washing out—being eliminated—from the UPT program. Instead of dreading this responsibility, as some people might have, I found I enjoyed flying with weaker students because I remembered well my own struggles as a student pilot. Being a fairly experienced IP by this point, I could often find a different way of showing or explaining how to fly various maneuvers, especially landings, in a way the student hadn't heard before and which might make more sense to him—or to *her* once women were trained to be pilots.

Sometimes, however, a student was just so lost he couldn't satisfactorily or safely complete the pilot training program, and he had to be eliminated. We called that a "flying training deficiency" or, informally, being "washed out of the program." It was sad for students and instructors… but it had to be done.

To this day, I still remember the first student I was forced to put up for an elimination flight. I really liked this lieutenant. He was young, energetic, and intelligent, but he could neither safely fly the airplane nor keep up with the demands and pace dictated by the training syllabus. After I recommended him for an elimination flight, and he subsequently failed it, I felt terrible. I felt even worse after meeting him and his wife and child at the base chapel when she tearfully told me he was being discharged from the Air Force and would have no job. All I could say to her was, "We saved his life. He could not safely finish pilot training." I totally believed that then, and I believe it now. But, even writing this now, decades later, it still saddens me.

On the brighter side, for every sad story, there are dozens of happier ones. I'm sure I could fill an entire book (or two) with stories of young student pilots I instructed during my years at various Air Force bases where I was either assigned or where I inspected. A few of my favorites from Reese AFB follow.

The Mustachioed Second Lieutenant

As a flight commander, I usually had one or two students in each of our classes that I would fly with regularly, in addition to occasionally flying with other students in the flight to sample how they were performing—and how their own instructors were teaching. One of my C flight students—we'll call him Lieutenant Winters to protect the innocent—was not doing well in his flying skills as he approached the point in the program where he'd be due for a contact check ride. To make matters worse, the Lieutenant had decided to grow a mustache.

To say that the Air Force discourages facial hair for any of its members would be an understatement, and to say that it doesn't want students to grow mustaches would be an even greater understatement. There is a saying that goes, "There are two kinds of mustaches in the military, those that look good and those that are within military standards, and you can't have both at the same time." But Lt. Winters wasn't outside regulations, so I couldn't tell him he had to shave it off.

As the Lt.'s check ride date approached and my dissatisfaction with his mustache—not to mention his flying—progressed, I made him an offer and, hopefully, gave him an extra incentive to do well on his check. Check rides were graded on a scale of excellent, good, fair, and unsatisfactory. "If you get an 'excellent' or 'good' on the check ride, you can keep the mustache," I told him. "If you receive an 'unsatisfactory,' the mustache will have to go." He agreed.

The next day, I didn't see Winters as he went to the squadron's Check Section to fly his evaluation with one of the designated check pilots. The following day, the Lt. reported to my office… still wearing his mustache, but only on the left side of his face. The right side was clean-shaven. As I stared at him, trying to make some sense of what I was seeing, he proceeded to say, as he stood at attention, "Sir, I got a 'fair' on my check ride." I tried not to laugh but couldn't stop myself as I sent him off to the shower room to shave the remaining half of the mustache. Lt. Winters successfully graduated from pilot training, and the last I heard, he was flying C-9 hospital planes.

I have a tendency to goof around probably more than I should, but at least I amuse myself, if no one else. Given my own difficulties with the UPT program while I was a student, I was probably more sympathetic to students who were having problems than other instructor pilots were. I wasn't above making jokes during debriefings following less-than-spectacular flights, if warranted, to keep the student's spirits up. I knew that sometimes a student could be harder on himself than necessary, and that could turn into a spiral of self-doubt, leading to further poor performance. Berating a student who already knows he performed poorly doesn't help him improve and has exactly the opposite effect. I learned that from personal experience as a student years earlier.

J.C. Penney Is Hiring

Captain Bob Watson was my assistant flight commander and a fellow Vietnam veteran, having also flown O-2s there. We had much more in common, though. Bob came home from Vietnam to a SAC assignment flying a large aircraft, in his case, a KC-135. He also managed to leave SAC for a T-37 instructor job. No wonder we became instant friends!

Bob would get a chuckle from something I would do on occasion to break a student's depression after an unsuccessful training flight. He and I still laugh about one of my techniques whenever we meet, even forty-plus years later.

As Bob recalls, I'd conduct the debriefing as professionally as possible, stressing some positives (if any) as well as the substandard performance. I never wanted to have a student leave my desk feeling hopeless. If I saw the student was being overly critical of his own performance, at the end of the debriefing, I'd kid around a little to break the tension and then, with all the flourish I could muster, reach into my desk and slap an application to J.C. Penney in front of him and say, "Hey, if all else fails, I hear they're hiring in the shoe department."

I'd then sign the application and say, Tell Penney's manager I sent you."

As the story spread, students came to expect this, and I'd swear they sometimes did poorly just to tell their buddies I'd handed them an application. I know they didn't do poorly on purpose, but I'll bet there are still some of those J.C. Penney applications, signed by me, in former students' desk drawers.

Each flight needed a good guy and a bad guy to deal with situations calling for a hardass or a softer touch. This wasn't written anywhere, but it seemed to work. Normally, the flight commander and the assistant flight commander would take turns being the good guy and the bad guy with each new class. Given my tendency to goof around and make jokes, I'd start a class of students trying to be the bad guy, but the students would see through me within a week or so. Poor Bob Watson would end up being the bad guy for almost every class we had, while I got to be the nice guy. Sorry, Bob.

During my time in Air Training Command, I often had an opportunity to fly with international exchange students. Countries with whom we were allies (at least at that time) sent members of their military to our country to be trained in many military skills, ranging from firefighting, maintenance, intelligence, and—in my case—pilot training. By and large, the international students we trained in undergraduate pilot training were well-screened by their countries, proficient in English, motivated, and intelligent. I said *by and large*. We also encountered a few international students who shouldn't have been sent to train in jets—or anything else, for that matter.

Sir, I've Never Driven a Car (but I want to fly a jet)

I remember telling an Ethiopian student who couldn't master taxiing the T-37, let alone flying it, "Think of taxiing as driving a car except use your feet while pushing the nose wheel steering button."

His response: "Sir, I have never driven a car."

Wow. I was speechless. I suddenly realized I was trying to teach someone who didn't know how to drive a car to fly a jet. But, to be fair, I later had the same thing happen to me when I tried to teach a Lt. from New York City to taxi the plane and fly, only to be told, "My family doesn't own a car, and I don't know how to drive."

In both cases, I sent the students to the base driver's education office.

The New Yorker made it through pilot training, but I don't recall what became of the Ethiopian. His country's government, which had sent him to the U.S., was overthrown, so his future was uncertain.

Mohammad and the Single Spin Recovery

Some of the international student stories have happier endings than others, and some have scary endings. No international students died on my watch, but I had a clear vision of my own passing with one of them.

As I mentioned, one of my duties as flight commander was flying with students who were having difficulties in the program and might have to be eliminated from training for what we called a flying training deficiency. In other words, it became obvious we were never going to get them to the proficiency level required to graduate within the time the syllabus allowed.

One such student was named Mohammad, one of many Iranian students we trained in the United States when the Shah was in power and Iran was an ally. This particular student had been extended in the program far beyond the amount of time normally allotted by the syllabus because his government was willing to pay for the extra training. It was becoming obvious, however, that Mohammad wasn't putting in the time and especially the effort to succeed in what is a difficult course of study. I was told to fly with Mohammed and brief our Operations Officer on my opinion of his chances of completing the program. This wasn't anything out of the ordinary for me, so I didn't think much about it.

Looking at Mohammad's training record, I could see he had a lot of extra training and had not learned the required procedures for the various maneuvers. The T-37 was the only aircraft in the Air Force inventory that was designed to intentionally spin, and there was a precise procedure to use in a spin recovery that would (theoretically) get you out of any spin you inadvertently got into in any aircraft (again, theoretically, and if the wings didn't tear off first).

The *Single Spin Recovery* was a seven-step procedure that had to be followed exactly in order to bring the T-37 out of a spin. Seven steps, 47 words, had to be memorized and followed exactly and quickly to recover. Not following these steps would not only not get you out of the spin but could further aggravate the spin and cause the plane to accelerate or flip over into an inverted, accelerated spin. Knowing the procedure and following it exactly was *absolutely* necessary and so ingrained in us that I can recite it today, many years later.

Well, I could recite it and perform the recovery, but Mohammad, as it turned out, definitely could not. After a few lackluster maneuvers and stall series in the flight training area, we had burned enough fuel to climb up to 25,000 feet to practice spins. I had Mohammad enter a spin and told him to recover. Recover we did not. He used an incorrect procedure and put us into an erect, accelerated spin, and we wrapped up into a tight spiral, turning one full revolution every three seconds while losing 550 feet per revolution. In other words, we were spinning and screaming at the ground at a fast rate. I quickly took control of the airplane and recovered.

As we climbed back up to 25,000 feet for another attempt at the spin recovery, I calmly asked Mohammad to recite the spin recovery procedure to me. He screwed it up royally. But, ever the professional instructor pilot, I told Mohammad to put us into another spin, and I'd tell him when to recover. He entered it incorrectly, and over onto our back we rolled, rotating rapidly, looking at the ground upside down as we spun toward the hard, cold earth. "Recover, Mohammad," I said. Recover, he didn't.

Now we're in an inverted accelerated spin, pinned against opposite sides of the aircraft. OK, I thought, this is where I can use this as a real teaching moment. I crossed my arms, looked at Mohammad, and said, "Recover the airplane, Mohammad, or we're both going to die."

Mohammed turned toward me—as the plane was spinning once every three seconds, losing 550 feet per revolution, and we were both being pinned to our respective sides of the aircraft—crossed his arms and defiantly said, "Mohammed not afraid to die!"

Arrgggghhhh! Well, I was *definitely* afraid we'd die, so I grabbed the stick, applied the spin recovery, called out of our area, returned to Reese, landed the aircraft myself, and told the Operations Officer that Mohammad should never touch an Air Force airplane again. After a final progress check with the Operations Officer, he didn't.

Flying with Royalty

In the 1960s and 1970s, the Shah of Iran was a valuable ally of the United States. He sent his son, Crown Prince Reza Pahlavi, to Reese AFB to train as a pilot, even though Reza had flown for years in Iran. He was young and bright. He was also eager to learn to fly, and his instructors were impressed with his attitude and skills.

My favorite memory of the Crown Prince of Iran, the future leader of his country—if it hadn't been overthrown—occurred in our squadron snack bar. Each flying squadron had a snack bar for the students and instructors to grab something to eat between flights since our schedules were too busy for us to leave the building to eat elsewhere. All students of the newest class in the T-37 program had to take turns, as their flying schedules allowed, pulling snack bar duty. The "snack bar officer" would sit at a small table near the door of the snack bar collecting small change for coffee, soft drinks, hot dogs, nachos, and doughnuts. (Pilots, obviously, didn't have the healthiest diets.)

At the time Reza was a student pilot at Reese, the Iranian revolution was starting to strengthen, and, as with all international dignitaries, there was a need for the Secret Service and Iranian security details to guard the prince. Since the prince was in our squadron for almost five months, we got to know—from a professional distance—many of the Secret Service agents assigned to the prince. As I said, Reza was a good guy, and he pulled his fair share of snack bar officer duty. I'll never get over how strange it seemed to be buying coffee and doughnuts from the Crown Prince of Iran while two Secret Service agents stood on either side of him with Uzi machine guns peeking out from under their suit coats.

Lt. Cline and Lt. Harris

Our unofficial policy was to give each student a "dollar ride" as a casual first-flight introduction to the aircraft. This was a chance for the student to try out the airplane, ask questions, and get to know their instructor. It also helped them get over any anxieties or misconceptions about flying the T-37. It was meant to be a fun experience.

My own introduction to Air Force Undergraduate Pilot Training (UPT) had been anything but fun. My initial instructor in the T-41 was a sadistic soul who rapped my knuckles with a short pointer and yelled at me at my least little mistake. As a result, he made me doubt my abilities and robbed me of the chance to develop a true love for flying. I didn't want that legacy to continue.

I knew from my own UPT days that student pilots had to work together to be successful. "Cooperate and graduate" was more than a catchy slogan. I was living proof that it worked. It was only after I started to draw on the support and guidance of classmates that I gained sufficient confidence to make it through the program.

Sometimes students opted for a basic straight and level flight. Other times, they wanted to wring out the plane and test their mettle. Usually, it was somewhere in-between. Sometimes a kid would overestimate the limits of their endurance and get sick, so we had an absolute rule that no one was *ever* to make fun of him. If a student still had problems a month into the program, that was a problem.

Lt. Cline didn't follow that rule. He was a "Zoomie" (AF Academy grad), and he had quite an ego. When his classmate Lt. Harris lost his breakfast in the plane, Cline was merciless. He teased and goaded him until I wouldn't have blamed Harris for beating the crap out of him. Unknown to them, I was seated in my office, overhearing the verbal abuse. When Harris remained silent, I interceded.

I stormed into the flight room and glared at the silent students.

"Capt. DeBruler," I bellowed at the flight scheduler, "who's flying with Lt. Cline tomorrow?"

"Capt. Buis," came the reply.

"Pull him," I growled as I fixed my glare on the now wide-eyed Cline. "I'm flying with Cline."

My eyes narrowed as I walked toward Lt. Cline and said, slowly and deliberately, "Eat a BIG breakfast tomorrow. When you get sick—and you WILL get sick—I don't want wimpy dry heaves. I want substance. Massive volume." I turned toward my office, then swung back around. "And don't bring some pathetic little air sickness bag. Bring a large plastic trash bag because you're going to need it."

Then I turned to the entire class, "Make no mistake about this. *Nobody* in this flight makes fun of another's misfortune. You're all in this together, and you can tease one another in fun all you want, but when something goes wrong, you stand together. And when Lt. Cline comes back tomorrow stinking of his breakfast, you give him the sympathy he should have shown Lt. Harris today."

I slept well that night. I suspect Cline did not.

The next day, we strapped in and taxied to the runway. "I've got the airplane, lieutenant," I said matter-of-factly. "You sit tight."

I took us up to 20,000 feet and began challenging the airplane to its limits. We pulled negative Gs, positive Gs, and transverse Gs—we were up and then down, first to one side, then to the other. I jerked the plane around and slammed Cline against the canopy rail, then barked orders to look out one window while I rolled sharply in the opposite direction. We dipped nose down, then nose up, then upside down, then right side up again,

Holy crap...is this kid ever going to blow lunch?

I tried to ignore the faint, unsettled rumbling in my own stomach as I pulled back on the stick and took us into a vertical stall, then jammed the stick forward and lifted the novice pilot out of his seat.

I pressed the plane into a series of barrel rolls, and finally, I heard that distinctive urping sound through the intercom as Cline filled his air sickness bag.

Success! Who knew victory could smell so gross? I flipped my oxygen regulator to 100% oxygen.

Then I leveled out the plane for the return home. "It happens to the best of us," I said, as Cline propped his head against the seatback and closed his eyes. "You guys are in this together, and in the future, you'd better remember that. The mistake or misfortune you laugh at today might just be yours tomorrow. I think I just proved that."

Back on the ground, I saw Cline heading in the opposite direction from the flight room. "Not so fast, Lt. Cline," I stopped him. "Walk slowly into the flight room, carrying that bag, and announce that you will never make fun of anyone again."

"Yes, Sir."

And he did. Then I told him, "Now go throw out that bag far, far away from here."

Yep, the Ejection Seats Work as Advertised

BOOM! The explosion was loud enough to be heard through my helmet and ear plugs and over the loud noise of the engine running off my left shoulder. The first thing I saw was thick smoke rising from two airplanes down the line and to my right in the chocks. Then, I spotted an ejection seat lying face down on the ramp behind one of the planes. *What the heck just happened?*

Each summer Air Force Reserve Officer Training Corps (AFROTC) programs from across the nation offer their senior cadets an opportunity to fly in an Air Force aircraft as an incentive to continue in the program and also as motivation to consider filing an application for pilot training. Most of these flights are conducted at pilot training bases, where the cadets get a chance to fly in either the T-37 or T-38. Many of the instructors I've known weren't too thrilled to fly with the cadets due to frequent bouts of air sickness and even cries of "I want to go back and land!" by the occasional cadet (and, occasionally, similar thoughts cross the mind of an instructor).

Personally, I enjoyed these orientation flights, and I became pretty successful in heading off potential airsickness and calming nerves. My secret: Let the cadet fly when it was safe to do so—obviously, when high above the ground—with copious amounts of praise. I'd get them so busy they forgot about being sick or nervous. This approach never failed me at Reese or at Randolph AFB.

The normal routine for preparing a group of cadets to fly with us was to hold three separate briefings on the procedures to be used in the three phases of the flight: pre-flight, actual flying, and post-flight. We focused heavily on safety and on emergency procedures, of course.

The first briefing informed the cadets about what would be happening during their visit to the squadron. The second was given by the enlisted life-support staff and *thoroughly* covered helmets, oxygen masks, and parachutes and then stressed the use of the ejection seats. Getting in and out of the plane safely wasn't as easy as it may sound, since we each took up two cadets, one at a time, during our sortie, and we switched students on the ramp as quickly as possible.

As part of this second briefing, dummy ejection seats were used by each cadet, who practiced getting into the seat, properly strapping in, removing and reinstalling the seat's safety pins, recognizing an ejection command, and squeezing the ejection triggers. The dummy seat was an actual ejection seat which had the explosive charges removed so the cadets could practice raising the arming handles and squeezing the ejection triggers in a realistic fashion.

The third briefing was given to the individual cadet by the instructor pilot flying their actual flight, thoroughly covering the three phases of the flight: pre-flight, actual flying, and post-flight. Once again, safety and emergency procedures were stressed. In all three briefings, the command to eject was made clear: Repeat the word *eject* three times. EJECT-EJECT-EJECT!

For all intents and purposes, anyone seated in an ejection seat was sitting on top of explosive charges intended to shoot the seat and occupant up and out of the aircraft in an emergency. There was also a strap running down the back of the seat, which we called a "butt snapper," that ignited, tightening the strap and throwing the occupant out of the seat after a second charge disengaged the lap and shoulder harnesses when clear of the plane.

While on the ground, the seats were "pinned" by elongated safety pins attached to long red "Remove Before Flight" streamers to prevent accidental discharge from inadvertently raising the seat's arming handles.

Finally, the first cadet would get into the jet, and then both engines would be started. Following a flight lasting about a half-hour, we'd land and taxi back into the chocks, where we'd shut down only the right engine (the side where the cadet was seated) to lessen the time on the ground and save fuel. With help from a crew chief, the student would then secure the ejection seat, unstrap, and climb out.

The second cadet would climb into the plane, strap into the ejection seat, hook up his or her oxygen mask and communications cord, and remove the ejection seat's safety pin, all with the help of the crew chief. Then the right engine would be restarted, and the second flight was ready to taxi to the runway. This was called a "hot turn" since one engine remained running during the entire cadet exchange.

At all times, from the first briefing until the cadets were safely back inside the squadron building after their flights, safety was *constantly* preached and monitored by the instructor pilot and also by everyone on the flight line. Potential safety issues were hammered into the cadets through each phase of their day, one of which was the need to follow the strictest safety procedures when getting into, while sitting in, and when getting out of the ejection seats.

I had just finished my first half-hour flight, shut down the right engine, ensured the cadet properly inserted the ejection seat safety pins, and was waiting for the crew chief to help the first cadet out and the second cadet into the plane when I heard the explosion.

I didn't know what had happened, but I knew it couldn't be good. I was uncertain whether to continue with my second flight or shut down and quickly get off the flight line.

After a discussion with the tower, we were told it was a ground accident and to continue the flights. As I taxied by the still-smoking aircraft, I saw the canopy lying on the ground and an ejection seat sitting on the ramp. Since I was taxiing and had planes following me, that's all I saw besides fire trucks and an ambulance rushing to the smoking T-37.

When I returned to the squadron building after the flight, I learned that a cadet had ejected herself from the airplane on purpose. *What the ___?* Capt. Ken Hamlin, the chief of T-37 Standardization/Evaluation for the Wing and part of the accident investigation team, later told me that this group of cadets was from Puerto Rico. The primary cause of the accident was the cadet had difficulty with English and misunderstood her pilot when he said, "OK, it's time to get out." She mistook that as an order to eject, raised the arming handles, and squeezed the triggers. As advertised, the seat ejected with her in it.

The T-37 ejection system is not designed to be operated until the airplane is at least 100 feet above the ground and flying at a minimum of 120 knots. Since the cadet ejected at zero feet and at zero knots, the ejection seat did not clear the tail of the airplane, and the parachute did not fully deploy. Both items probably helped her to escape with only a black eye. The seat "butt snapper" did fire, and the lap belt disconnected, so she did not come down with the heavy seat on top of her and crush her, and the fact that the parachute didn't fully deploy cushioned her fall when she landed on her back on the concrete ramp. It's also fortunate the crew chief hadn't yet had time to reach into the cockpit to help install seat pins, or he'd have been hit in the head by the seat and possibly been killed.

As an aside, Ken told me the accident investigation also revealed that the ejection seats did not have enough thrust for a heavier occupant to clear the tail of the plane, even at 120 knots and 100 feet above the ground. I guess I was lucky I weighed only 165 pounds.

Finally, Water Survival Training—while living in the Texas desert

For most of my flying career, I'd flown over water for one reason or another. My first true piloting assignment was flying O-2As at Shaw AFB in South Carolina. We routinely flew over lakes and, sometimes, along the Atlantic coast. We also worked with the fighters at MacDill AFB near Tampa, Florida, and flew over firing ranges in the Gulf of Mexico.

My next assignment, also flying O-2As, was in Vietnam, where we flew to our areas of operation over the South China Sea, and our preferred bailout area was in that same water.

Next, I flew B-52s and crossed all kinds of water, ranging from the Great Lakes to both the Atlantic and Pacific Oceans. Not once during all of those assignments did the all-knowing powers within the Air Force see fit to send me to water survival training. Not once. It was never mentioned—until I moved to the desert, of course.

So, I get assigned to Reese AFB in Lubbock, in the heart of the high plains of West Texas, a semi-arid, almost desert-like environment with virtually no bodies of water. Only then did the all-knowing gods of knowledge in the Air Force personnel system deem it necessary to send me to water survival training. Only then, when there was no chance of my flying over anything containing more water than a large swimming pool, did the Air Force worry about my chances of bailing out and needing to know how to survive in a hostile water environment.

And, so it was that I ended up at Homestead AFB, near Miami, Florida, the home of USAF aircrew water survival school. And, of course, I ended up going to water survival training during the only cold period of the year in Florida.

The school itself was sort of like a vacation, other than the aforementioned cold weather. The classroom training was easy and actually interesting. Who knew you could make a set of water wings out of your flight suit? The actual water training wasn't quite as much fun—I've never liked swimming or anything much to do with water—but it wasn't difficult.

We were trained on how to disconnect our parachute while being drug though the water by strong winds after a bailout. They did this by pulling us, one at a time, behind a speedboat while we were still wearing our helmet and oxygen mask, with our parachute harness attached to the speeding boat by a long rope.

We learned how to deploy the survival raft and other emergency supplies in our ejection seat pack by being pulled into the air by another speedboat while we were attached to a parasail. After the boat disconnected me, I had about one minute, as I floated downward toward the Atlantic Ocean, to deploy my seat pack and blow up my water wings. Not too bad, and I enjoyed the parasailing.

I should say that I enjoyed parasailing exercises until what amounted to the final exam. On one of our last days, the staff had a practical exam designed to combine the various training components into one semi-realistic exercise. All the students were taken out into the Atlantic about a half mile from the Florida coast on something resembling a big flat-topped raft or a mini aircraft carrier.

We were equipped with the survival gear we would fly with if we were, in fact, doing an over-water flight, although I was never issued any of it in Vietnam other than water wings. We had flight suits, helmets, uninflated water wings, and a parachute survival pack strapped to our rear ends, just like it would be if we had ejected out of an airplane. The essence of the drill was for each of us to be pulled up into the air from the flat-topped mini-carrier and pulled out into the ocean before being released to parasail into the water. As we descended into the water, we were to deploy the seat kit, releasing the survival gear—including a small one-person life raft which would, hopefully, inflate before we hit the water. We were also taught to inflate our water wings as we descended.

After we landed in the water, we were to climb into the raft and float around in the ocean until a rescue helicopter came to pick us up. Small detail: We had to signal the helicopter using all of the signaling techniques we'd learned since the class was scattered in a relatively large area in the Atlantic.

Oh, yeah, did I mention that it was cold on the day we were dropped into the ocean? Accordingly, we were issued rubber suits—referred to as "poopy suits"—which fit over our flight suits and would, theoretically, keep us from getting hypothermia in the water during our five-hour float in the rubber raft. As luck would have it, the poopy suits weren't in the best condition, and mine, I was to discover later, had a few tiny rips and holes in it. That would present a problem.

As I was pulled into the air by the power boat from the mini-carrier and accelerated to about twenty knots in the parasail, my poopy suit began to fill with air through those little rips and holes as the air rushed by. By the time I was released, my poopy suit was so full of air that I felt as if I were in a balloon. I did a fantastic job of deploying the survival kit, ensuring the raft had inflated, and pulling the compressed air cylinders to inflate my water wings, but I was still in a poopy suit full of air.

I did a good job of releasing my parachute connects when I hit the water, but, because the poopy suit was full of air, I just bobbed around on my back like the balloon I had become. Not good. Not good at all. I couldn't get my feet or body into the water to swim to the raft.

Just when I thought *Things can't get any worse*, things got worse. The air in the suit started escaping, only to be replaced with salt water now entering. As it began to fill the suit, I made a quick swim to the raft and threw my arms over the edge, only to find that the bottom part of the poopy suit was too heavy to allow me to pull myself into the raft. I was becoming a sinker, slowly filling with water, and there was the real potential for me to become so heavy I'd sink completely, with or without water wings.

If there was ever a time to panic, this was it. Luckily, I tend to stay fairly calm in an emergency and figured if I could slowly pull myself into the raft a few inches at a time, I might be able to get the water to drain out some of the holes as I rose above them.

Slowly, over the span of what I estimated to be an hour, but was probably much shorter, I was able to pull myself over the edge of the raft a little at a time. Once I pulled my upper torso into the raft, I could feel the water level inside the suit going down. When I finally got my entire body into the raft, I was exhausted, but I didn't rest until I got out of that worthless poopy suit completely. I feared that if I somehow ended up back in the water, I would be so tired I wouldn't be able to struggle back into the raft a second time.

Emergency averted, I laid back in my little raft and actually enjoyed bobbing up and down in the ocean until that evening when the helicopter appeared, and it was time to signal where we were for pick up. Once the helicopter crew located me,

I let the cable and harness on the helicopter's hoisting mechanism hit the water to dissipate any possible static electricity and enjoyed the ride up. I had survived water survival school and could return to Reese AFB…and, once again, fly with no water anywhere to be found.

Up until now, I've referred to all of the student pilots—and pilots anywhere in the Air Force—as *he*, *him*, or *the guys*. Other than some Women's Airforce Service Pilots (WASPs) during World War II, all pilots in the military were males. That all changed in the mid-1970s. Beginning In 1974 in the Navy and Army and in 1976 in the Air Force, women could be trained to become military pilots.

The Air Force opened admission to both the Air Force Academy and UPT to women in the same year. The first women to graduate from Air Force pilot training graduated from Williams AFB in Arizona in 1977. Reese AFB was one of the next UPT bases to include women in their classes, and it came with mixed feelings from some of the male instructors.

Before the first female student pilots arrived at Reese, there was a thorough "scrubbing" of all buildings—offices, desktops, walls, and any other place that might harbor material the women could find objectionable. This might not seem like a big deal in this day and age, but in the previously totally male environment of a pilot training base, it required some legitimate removal of photos, pin-ups, jokes, and so forth. Of course, in a typical military/governmental overreaction, some supervisors went overboard, and it created unnecessary friction.

Outside most flight rooms, there was a framed board with a photo of each instructor and his short biography. In C flight, my flight at the time, we had each IP's photo taken in front of the then-iconic poster of Farrah Fawcett in her red bathing suit and big, blonde hair. We had to remove all of the instructor photos on the off-chance that someone would be offended. Poor Farrah. Poor C flight.

The women arrived, and, as it turned out, it was not a big deal. The Air Force survived. In fact, since I later had a daughter become an Air Force lieutenant, I'd say the Air Force is a better place for the changes.

And the female student pilots turned out to be pretty much as good or as bad as their male counterparts. There were some really good pilots and some not-too-good pilots, just like the guys.

After just two years on the flight line, I was selected for the position of Chief of the Student Branch. This was a career advancement usually reserved for a more senior officer, and while I loved being an instructor and thoroughly enjoyed working alongside my fellow officers in the 35th Flying Training Squadron, I eagerly moved the two blocks from the flight line to the 64th Student Squadron building.

As Chief of the Student Branch, I was responsible for all 300 students on base, and I developed and monitored the administration of policies pertaining to the discipline, physical and mental well-being, and general welfare of those students. I addressed each new class, explaining those policies and procedures.

I had eight class commanders who worked for me and were each responsible for supervising one class. The class commanders and I retained our positions as instructor pilots, and while the class commanders normally flew only with students in their respective classes, I had the privilege of flying with whatever flight I wanted to. For the first time in more than two years, I actually got a chance to eat lunch at the Officers' Club—a real step up from the flight line snack bar!

Once again, I was fortunate to have a great squadron commander, Lt. Col. Barry Brower, who was not only my commander but a great mentor. My job was relatively easy due to the high-quality, hand-picked officers who were my class commanders. They performed their jobs well, and I just made certain they had the support they needed.

We did have one not-so-great duty in the student branch: handling any ceremonial duties the wing commander or any other full colonel on base couldn't find anyone else to handle. As a result, we became the base experts on organizing every pilot training class graduation ceremony, every formal graduation dinner (known as a *dining-in*), and just about any other ceremony one could think of on the base. It often seemed a pain at the time, but, as it turned out, that experience came in handy in my later Air

Force positions and also in my civilian career to follow. Who knew?

One position that is a definite plus in any pilot's career progression is serving time in what is known as the Standardization/Evaluation section, abbreviated as Stan/Eval.

Standardization/Evaluation does what the title implies. These pilots ensure that all of the pilots within a unit (in this case, the entire wing at Reese AFB) are flying and instructing in the same way and according to all of the manuals and regulations—in other words, flying within all of the standards of Air Training Command and the Air Force. The evaluation portion of the job description means that the Standardization/Evaluation section does the evaluation—instructor pilot check rides and paperwork—for the wing.

Each pilot in the Air Force is given a number of "check rides" each year in various skills, depending on the type of aircraft they are flying. The phases that are evaluated in ATC were contact (flying looking out of the cockpit windscreen), instruments (flying in conditions where you cannot see outside the cockpit), formation (flying within three feet of another aircraft), and navigation (flying from one place to another without getting lost). Each instructor pilot was given a check ride each year in at least two of these phases, contact and instruments, and was required to pass those check rides to maintain their instructor and pilot status.

After spending less than a year in the student branch chief position, I was selected to become a member of the wing's Standardization/Evaluation division. While it meant that I would be the one conducting check rides for other pilots, it wasn't an easy way out of work. Now I was expected to be the expert in all things instructional and in all things flying the T-37. I still got to fly some student instructional flights, but now most of my sorties were flying with other instructors and evaluating them. We also evaluated the Lt. Colonels and Colonels who remained current in the T-37. I might mention that other pilots weren't too fond of the evaluators in Stan/Eval.

Not long after I arrived in Stan/Eval, T-37 branch chief Ken Hamlin left, and I moved into the position of Chief, T-37 Standardization/Evaluation. Now I supervised all of the T-37

Stan/Eval evaluators. Have I mentioned how much more fun I was having in ATC than in SAC?

Other than the Lt. Colonels and Colonels at Reese, the only persons any of us in Stan/Eval feared were Headquarters Air Training Command Standardization/Evaluation (ATC Stan/Eval) team members. These were six to eight instructor pilots who had been hand-selected to thoroughly inspect all of the UPT bases in the country and give each base a rating, a rating that could potentially make or break the careers of everyone from the lowliest IP to the Wing Commander. We weren't too crazy about the ATC Inspector General team, either, but, at that time, those guys looked only at paperwork and didn't give check rides. Far less stressful inspections!

In the fall of 1978, ATC Stan/Eval arrived at Reese AFB for our annual inspection, and one place the team always zeroed in on was the base Stan/Eval division. They rightfully needed to know that the pilots evaluating the base pilots knew exactly what they were doing and were grading accordingly. Plus, they usually gave each base Stan/Eval member a lengthy check ride and intense ground evaluation. As the new branch chief, I knew I had an excellent chance to go under the microscope. And, of course, I did.

My evaluation with one of four T-37 command flight examiners began with a long and, to me, painful ground evaluation which covered more aspects of flying the T-37 aircraft—maneuver parameters, operating limits, Air Force and federal flight regulations, and emergency procedures—than I would have thought possible. I believed I had done alright on that, but a major part of the ground and flight evaluation was just getting started. We briefed the flight in detail, including in which parts the evaluator would be playing the role of a student needing correction and instruction and in which parts I'd fly the aircraft myself. I wasn't used to being the guy under the microscope, and I didn't sleep well the night before.

My check flight went well from my biased point of view, but the actual debrief would let me know how the examiner viewed it. The debriefing was all business on the part of the examiner, but I could sense that he was starting to lose some of his strictly business façade as the briefing progressed. At the end, he shook my hand and told me I had done an outstanding job both on the ground and during the flight portions. I could breathe again; my career wasn't over. As a matter of fact, I received the highest possible rating on the check ride, and, as it turned out, I had taken a large step toward my next job. But I'm getting ahead of myself.

While my Air Force career was rolling along at Reese, Karen decided to put her free time to good use now that she had given up her teaching position to follow me to Texas. She took the Graduate Record Exam and subsequently enrolled in a Master of Curriculum and Instruction program in the College of Education at Texas Tech University. Her theory was she should keep her classes general, instead of specializing in a particular subject or in administration, so as to make it easier for her to qualify for more positions in the future as we continued to move.

Karen found out in her first semester at TTU that master-level classes never began before 4:00 in the afternoon since most candidates were currently teaching. A drawback of that was it would take longer to complete the degree requirements. A positive was she was available to fill part-time research assistant positions for professors in the College of Education and still have time for active participation in Officers' Wives Club activities, including volunteering weekly at the base thrift shop—and sewing new unit patches on my flight suits and flight jackets every few months as I changed units on base. All this kept her busy and offered numerous opportunities to make close friends, both socially and professionally. (Well, maybe not the patch sewing.)

Meanwhile, Lubbock's housing market was booming, especially in the southwest part of town where we resided, and a new development appeared at the west end of our street. The house we bought upon our arrival in Lubbock was nice, but a visit to the new home of a fellow instructor pilot and his realtor wife got us thinking maybe we could build something even nicer. This time we'd have a custom home built just for us.

Since we thought we'd be in Lubbock for two or even three more years, and neither Karen nor I was eager to leave town anytime soon, we decided to sell our first house and build a new one. We selected a prime lot and worked closely with a builder to plan every aspect—floor plan, exterior appearance, and every item, from floor to ceiling inside. The house included a sunken living room, a separate family room with a bar, custom cabinets, and a covered patio. Karen enrolled in an evening class at the university to enable her to create a landscaping plan.

While the home was under construction, Karen completed her Master of Education degree and scheduled an interview for a classroom teaching position with the Lubbock public schools. Things went well until the superintendent asked her, "Do you speak Spanish?"

"No."

Then he told her, "We like to hire people who have your experience for our poorer schools."

Karen knew little about the public schools in Lubbock, but those sounded like two red flags to her. She'd have to think about her options.

Less than a week later, out of the blue, Karen received a phone call from the chair of the Department of Early Childhood Education at Texas Tech asking her if she was interested in a part-time instructor position teaching two methods courses, math and science, to undergraduates at the university. She'd been recommended by the two professors for whom she'd conducted research while a graduate student. After a short interview, she got the job and was to start in September of 1978, the fall semester.

The new house was completed that summer, and we quickly sold the first house. Then, with the help of my buddy, Captain Ken Broda, we moved the three blocks from that first house into the new home in July of 1978 and began to install the landscaping and finalize an order for custom draperies. The sun was shining, the birds were singing, and things couldn't have been more cheerful, except…

This is where the story of the new house, the story about Karen's teaching position at Texas Tech, and my outstanding check ride during the ATC Stan/Eval inspection all intersect. Or *collide*, depending or whether Karen is telling the story or I am.

Let's start with the fact that a position in Headquarters ATC Stan/Eval was a highly sought-after and career-enhancing job. There were only three or four ATC Command Flight Examiners each in the T-37 and T-38, and they were by-name assignments. That means they were not open to the normal assignment process. The individuals in those jobs were hand-picked from all of the T-37 and T-38 instructor pilots at all of the bases in Air Training Command.

As a result of my record at Reese AFB, my performance on my ATC Stan/Eval check ride, my wide experience level in various jobs within the pilot training program—and, undoubtedly, recommendations from Lt. Col. Brower, Col. Bagley, Lt. Col. Bradshaw, Maj. Neff, and others at Reese—I received a phone call informing me that I'd been recommended to become a member of the T-37 team in Headquarters ATC Stan/Eval at Randolph AFB. The job was mine if I agreed to it. I'd have to be crazy to turn down this opportunity.

"Think about it, and give us a callback," I was told.

Karen came home from her <u>very</u> <u>first</u> day teaching classes at Texas Tech, all smiles as she enthusiastically told me about her day. "This is great! So much better than teaching elementary! No discipline issues. No lunch counts. No recess duty. The students actually listened for an hour and a half and asked great questions…

"This is what I want to do from now on," she beamed.

Now what would I do? I was silent for a moment as Karen set down her tote bag before informing her, "I got a call today from ATC Stan/Eval. They want me to come to Randolph."

Silence. Then, "When?"

"As soon as possible."

Much longer silence. Then, looking me squarely in the eyes, she said, "*You're* moving. I'm going to have to think about this one."

Unfortunately, I was certain Karen really meant that. How could I do this to her? How could I do this to *us*? But how could I turn down this offer?

We'd been in our dream home for exactly six weeks. Didn't even have any trees yet, since the nursery wouldn't plant the live oaks and sycamores we'd purchased until a month ending in R. Custom draperies were still on order. The grass was just beginning to fill in.

Karen and I *both* thought about it. She realized by the next day that this is what we'd both signed on for as a career military officer and as a spouse of one. We would make the move to San Antonio. Karen's only request was, "Can you hold off until I finish the semester?"

So I did.

I called Colonel Bill Keller at ATC Stan/Eval, accepted the job, and scheduled the move for the end of January. Karen and I made a quick trip during Thanksgiving break to select a new house near Randolph AFB, and we easily sold our Lubbock home. After all, who wouldn't want to buy a new custom-built home without undergoing the hassles of construction and landscaping?

By the end of January 1979, Karen and I said farewell to our friends in Lubbock and at Reese AFB, and Karen, Barney Dog, and I waved good-bye as we headed back to San Antonio and Randolph AFB. We had absolutely enjoyed the assignment at Reese and Lubbock, but it was time to turn the page and see what new and exciting things lay ahead of us at ATC headquarters.

A T-37 "Tweet" inflight over West Texas. Mike is the instructor in the right seat—three feet of spacing from the leader.

Author's note:

My appreciation for Karen's—and, later, my entire family's—sacrifices for my military career is immense and something I cannot adequately put into words or repay. Karen sacrificing for my career in this chapter will be repeated often, and my children will also sacrifice. It's something all military families are faced with, whether that is a four-year enlistment or a twenty- or thirty-year career. Spouses repeatedly leave friends, jobs, and homes they love on behalf of the military member. Veterans' families don't get paid, and they don't get free food from restaurants on Veterans Day, but they deserve a special holiday all to themselves.

As much as I admire and respect our military, I admire and respect the sacrifices their families make on its behalf. God bless Karen and our daughters, Lori and Katie, and God bless all those who sacrifice a "normal" life on behalf of our nation. They are *my* heroes!

Chapter 5

Top Rung of the Ladder

At the end of January 1979, Karen, Barney Dog, and I arrived at Randolph AFB in Universal City, Texas, near San Antonio, for the second time. I was now a brand-new member of the Air Training Command Headquarters and one of only five T-37 instructor pilots in an elite unit called ATC Stan/Eval, Air Force speak for Standardization and Evaluation. *I was hand-picked for this?* Amazing but true.

Karen and I had purchased a two-story Ray Ellison home in November. It was newly constructed and just a few miles from Randolph. Anyone who lived in the San Antonio area in the 1970s or 1980s has to remember Ray Ellison Homes since they created new developments across the area built on the fringes of the city, mostly to the northwest and northeast. The homes weren't custom-built, but the company enabled the buyer to select flooring and wallpaper. A bonus of working with such a huge developer was they had crews available for every possible phase of construction and repair, if needed. Since Karen was now unemployed, she took on the task of making the house our home while I educated myself on the new job.

Our office of T-37 and T-38 instructors and evaluators was responsible for ensuring *all* of the Undergraduate Pilot Training wings across the United States, the Undergraduate Navigator Training program at Mather AFB in California, and the flight screening operations at Hondo, Texas, and at the Air Force Academy were conducting training in accordance with appropriate Air Force and ATC regulations, guidance, and procedures. Quite a mouthful! This was an important—but exciting—responsibility.

Arriving at the office on the first day, I was put to work immediately. There was no slow introduction to ATC Stan/Eval, as we were expected to already know what we were doing and how UPT and wing-level evaluations operated. I did have to do the normal pilot checkout in the local area as part of my familiarization, but I had been assigned to Randolph as a PIT student and had been back many times on cross-country flights with my own students, so that was quick and easy.

I had reached the top of the ATC instructor-pilot ladder. We inspected all ATC UPT bases and their pilots, which was a busy job. We traveled to a different UPT base—from Arizona to Georgia—every couple of weeks, spending days at each base to inspect all aspects of the program—paperwork, classes, and flying—and then debrief.

I found the officers, enlisted members, and civilians assigned to Headquarters, Air Training Command to be highly competent and also quite personable. The commander of ATC, General John Roberts, was a four-star general who worked right down the hall from our office. Shortly after I arrived at Randolph, General Roberts was succeeded by General Bennie Davis, who was a highly-respected leader who always said hello to me as I passed him in the hall.

The ATC Deputy Chief of Staff for Operations for our section was a one-star general, Brigadier General Kenneth Fleenor, who had been a prisoner of war in Vietnam. He occupied the office next to ours. This proximity to such high-level officers was new to me.

The people I worked with quickly became good friends, and I enjoyed going to work. The commander of our unit, Colonel Bill Keller, was a nice guy whom I had met a few years before. I considered him to be a friend even before I arrived, as well as now my boss. My immediate supervisor was Major Ron Mastin, for whom I enjoyed working due to his fine sense of humor and his integrity. Ron had been a Vietnam POW with General Fleenor.

The other T-37 pilots in the office included Captains Bill Blackmon, BJ Cote, Dave Taylor, and, later, Loren Stendahl, and they all made the job and the many temporary duties (TDYs) away from home more enjoyable. I still keep in touch with many of them.

The ATC Stan/Eval job itself was a unique environment in that we didn't have the threat of inspections of *our* operation to worry about. No one inspected us; we gave check rides to each other. No fear in that—just fun.

I quickly realized I had reached the top of the T-37 pile when, after I was told of my additional duty as the publications officer, I mentioned, "I don't know anything about maintaining publications! We could be in big trouble when we're inspected."

My new buddies in the office looked at each other, started laughing, and said, "Who's going to be inspecting us?" I paused for a moment. Short answer: No one. We weren't subject to *any* inspections. I thought *This job is really going to be fun.*

The only thing I knew about publications when I started the job was how to keep my own T-37 Dash-One (the bible for how the T-37 operates, its flight and instrument limits, and normal and emergency procedures which we largely memorized) current. The Dash-One often needed changes that required inserting new, revised pages for old ones or what we called "pen and ink changes" that required scratching out various lines in the manual and writing new ones. Beyond that, I knew absolutely nothing about maintaining the vast library of Air Force and ATC manuals and regulations filling the shelves lining our office walls. But I'd learn, although I can't say I ever truly enjoyed *that* duty.

As the publications officer, I became the person responsible for writing and updating Air Force Manual 51-4, *Primary Flying, Jet*, the "How to Fly the T-37" publication. It gave instructions on such basics as the correct airspeeds for various traffic pattern configurations to more difficult instructions, such as how to fly in emergency situations. Oh, if only my instructor from my days as a struggling student pilot in T-41s could see me now!

One of the many things I liked about being stationed at Randolph, and serving in the military in general, was that, after I'd been in for a few years, I'd run into people with whom I'd been stationed previously. As I said, I knew Col. Keller, but I also knew the other pilots in Stan/Eval from when they'd inspected our squadron at Reese, and I often ran into people I knew while elsewhere around Randolph. Plus, it was like a reunion going from base to base for our inspections. The longer you were in the Air Force, the more people you knew—and you were never alone at any base after a seemingly short time.

For example, my assistant flight commander at Reese AFB, Bob Watson, ended up working in the headquarters building on the floor below mine. The friendships you make in the military are a large part of the military experience—one of the best parts, for sure.

The same circle of friendships extended to the military member's spouse and family. Case in point: Karen had been good friends with Bob Watson's wife, Cindy, and when Bob transferred to Randolph, Karen and Cindy's friendship picked up where it had left off in Lubbock.

The same warm welcome I received from the men in my office extended to my family. At an early social gathering, Karen met a wife who was teaching at Coronado Village Elementary School in Universal City. Karen mentioned that she'd applied for a position in that school district. The kind wife put in a good word for Karen, and she was offered a third-grade position for the fall at the same school. The power of networking!

It was good for Karen to have a job to challenge her and also to make friends with whom she could spend time outside of work since the Stan/Eval teams, as mentioned above, spent much of their time on the road (or in the air). She formed a close bond with a fellow teacher, Carolyn, who remains a friend to this day.

They worked closely in an open-classroom setting on weekdays and often met at the school on Sunday afternoons to change bulletin boards and arrange materials for the week.

I was forming lasting friendships, too. On our inspection trips, we sometimes had two-room suites, and Bill Blackmon and I usually roomed together. We had a similar sense of humor and got along great. We were always on the lookout for ways to have fun, and the opportunities in a flying unit, no matter what level the unit, were endless.

When a new captain showed up for the navigator side of our unit, I saw a chance to "welcome" him to the office. I mentioned to Bill, "I'm cold. Do you want me to turn up the heat?" Bill knew as well as I did that there were no heating controls in the office, and the only thing remotely connected to the building's furnace was a thermostat which showed the current room temperature, nothing more.

I made certain the "newbie" saw me as I walked over to the thermostat and said into it, "Turn up the heat two degrees." Bill and I both knew this would have absolutely no effect on anything.

For the next few weeks, whenever we saw the new guy walk by our area, either Bill or I would ask him to turn the heat up or down. He'd go over to the thermostat and speak into it, saying either, "Turn up the heat" or "Turn down the heat." The rest of the guys in the office would snicker as they caught onto the joke.

It wasn't until our secretary, Joanie, tipped our new guy off to the fact that he was talking to a mere thermostat that he wised up and realized he'd been had. He later tried the same trick on the next new guy, but, evidently, he wasn't nearly as convincing as Bill and I had been.

One way Bill and I differed was in our neatness standards. I had always been a relatively neat and organized person, probably due to my mom's insistence that my siblings and I keep the house and our rooms neat, since she had no tolerance for messes with five kids. Bill, on the other hand, had an entirely different approach to "organizing," which manifested itself on the top of his desk and in his unpacking while on inspection trips.

Whereas I'd unpack my B-4 bag (a large, military-issued, zippered suitcase) carefully into drawers as soon as we reached our rooms, Bill's idea of *unpacking* was to unzip all three sides of the bag and slide it across the room so his clothes spread across the floor, where they would remain for the duration of the trip. I will admit his method was faster.

By the way, keeping my office desktop clean and orderly would come at a cost and cause me to stop appearing quite so neat. I overheard Colonel Keller's replacement comment, "Give that job to Jackson since his desk is so neat." From that point on, I made sure to keep things spread out over the desktop so he'd know I was busy.

While we were at Bill's house for an office social gathering one evening, Karen mentioned to Bill and his wife, Sherry, "Water heats faster if it's cold than when it's hot." (She meant that cold water changes temperature at a faster rate than hot water does.) To this day, neither Karen nor I have any idea why she said that or remembers what we were even talking about, but consider the fact that alcohol was usually served at such gatherings. At any rate, Karen must've sounded convincing because no one disputed her statement.

On a number of occasions on our future trips, when Bill was making instant coffee, I'd wait until his water was almost ready to boil, and, with great ceremony, I'd dump his water down the drain and refill the pan with cold tap water. He knew what I was going to say. "I was helping to speed things up for you because Karen says, 'Cold water heats faster than warm water.'" He thought it was kind of funny…the first few times.

Each of the members of our unit had a unique personality and unique experiences. BJ was the unit's only bachelor, and he always had a funny story about things that happened to him the previous night or weekend. Dave "Guff" Taylor was our southern "good ole boy," and since most of the pilot training bases were in the South, we relied on him to suggest the best places to eat off-base in the evening. For example, Guff was about the only one among us northern boys who had ever heard of *Krystal's*, a fast-food hamburger chain. Guff loved the place and ate there whenever the opportunity arose on our trips to other towns since there were none in San Antonio.

Prior to my arrival at the ATC Stan/Eval office, the team was viewed as a "high threat." Visits were made without notice to the wings, and a difficult general knowledge test was given to and feared by all. Thankfully, the philosophy changed a year or so before my arrival from a fearful inspection to the "We're here to help" mentality. Formal and informal visits were scheduled a year in advance, with formal inspections conducted at 18-month intervals and an inspection of only the wing Stan/Eval section at nine months in-between. A general knowledge test was based strictly on need-to-know material. I'm glad I arrived after the change. The purpose of *our* visit was to provide the wing commander, the deputy commander for operations, the squadron commanders, and ATC Headquarters with a report card.

We spent seven busy days at each UPT base inspecting and evaluating everything under the sun that was connected with the base's flying and training program. Each team member conducted training flights with students and check rides with instructors. In addition, we each had our specialty.

We usually departed Randolph AFB on a Monday and flew in either a T-43 Air Force transport or a commercial airliner. The exception was when we inspected Laughlin AFB in Del Rio, Texas, when we car-pooled. When we arrived at the inspection site, we were given our local flying publications and an airspace briefing, and then we were off to the BOQ to review the next day's schedule.

On Tuesday and Wednesday, we flew student sorties and gave check rides to the instructors. On Thursday, we focused on our specialties, mine being the operations of the student squadron since I'd spent time in that unit at Reese. I examined the records of incidents involving discipline issues with students and their resolution. I spoke to student pilots, class commanders, and the student squadron commander to ensure theirs was a well-oiled operation.

Meanwhile, the other members of our team checked to see if all publications were up-to-date, if the classroom syllabus was followed for each class, if both instructional rides and check rides were conducted according to ATC standards... In general, we looked at *all* facets of the UPT program. It was intense.

Each evening, the team would gather to share information on what we'd found. There were always areas someone felt we should examine further, and the meetings lasted into the night as we planned the next day's activities.

On Friday, we inspected the wing's Stan/Eval unit and flew flight examiner check rides. Saturday was devoted to student grade book checks. On Sunday, we wrote an extensive formal report. The rating system was Excellent, Satisfactory, Marginal, and Unsatisfactory. The overall ratings given were usually excellent or satisfactory, although sub-areas, such as grade books, could be rated marginal.

On Monday morning, we gave a shortened out-brief to the flight commanders, followed by a formal presentation in a stand-up debriefing to the wing commander, the wing's deputy commander for operations, the two flying squadron commanders (T-37 and T-38), the student squadron commander, all of the wing's Stan/Eval team, and the staff of all of those. We included slides to document our findings. This was a big deal, since we could praise or criticize so many facets of their program. Then we flew home.

As if the frequent trips to other bases didn't keep me busy enough, I doubled my workload –and stress level—by beginning work on my Master of Business Administration degree from Webster University. Webster is located near St. Louis, but all of my classes were held in the evenings at nearby Fort Sam Houston. Several colleges and universities offered courses at bases around the country since the military encourages continuing education to improve one's knowledge and skills.

I honestly didn't want an MBA, but I knew it was one of the items—"square fillers"—that would set me apart from others when it came time to be considered for the next promotion. Every class I missed due to being out of town on an inspection trip normally required me to write a five-page research paper, not to mention missing the class lectures and discussions that would be included on the next exam.

About six months prior to my departure from ATC Stan/Eval, Captain Loren Stendahl was selected to take BJ Cote's position on our team. Loren and I knew each other from a Stan/Evan inspection of his base, Columbus AFB in Mississippi, and I sensed we would get along well. So, I was happy to give him the orientation rides and the required local check-out training flights in the local flying area, both of which were necessary before joining our inspection team.

When we went up on a contact flight in the Randolph airspace, he and I took turns flying and evaluating each other as we changed roles as instructors and evaluators. We did the required check-out items, such as showing him the climb and descent corridors, local flying area landmarks, traffic pattern entry points, and other area procedures. Loren knew he had reached the top of the evaluator ladder when, after we landed, I had him fill out the paperwork for *my* annual instrument check ride, even though we hadn't flown any instrument maneuvers other than one localizer approach into Kelly AFB.

Then, when he asked me when we'd debrief the overall familiarization flight, I looked at him as if he were nuts and said, "Welcome to ATC Stan/Eval, Captain Stendahl!" He knew he'd found a home.

In March of 1980, I was promoted to major and was among the top ten percent of those promoted. Subsequently, I was selected to attend Air Command and Staff College (ACSC) or some other in-residence field-grade professional school as my next assignment. Life was continuing to be very good, but…

As with all jobs, military or civilian, there were some good and some not-so-good facets. This job was nearly all good. I was in a top position within ATC, and I enjoyed flying the T-37. What I didn't enjoy was how much time I was required to be out of town during our inspections of the pilot training bases and the Air Force Academy. It averaged about ten days a month, and that quickly got old to me. It wasn't nearly as bad as the weeks I spent on alert when I flew B-52s because I was enjoying what I was doing, but I wanted to be home more.

Despite being a pilot, I've never really enjoyed traveling. At least, not overnight travel. That probably strikes many pilots as incongruous. Flying usually means traveling from one place to another, doesn't it? Well, I'm basically a homebody, and I don't like living out of a suitcase. Even today, I'm not that excited about going on a vacation.

When we found out in early 1981 that Karen and I were going to have a baby, my desire to find a different flying job at Randolph increased. As luck would have it, Lt. Col. Jack Rogers, my good buddy from ATC Stan/Eval, had been given command of the 559th Pilot Instructor Training (PIT) Squadron. I shared my desire with him, and he put in a by-name request to have me reassigned as a section commander in his squadron. I subsequently left my lofty position at the ATC headquarters and moved my office across the base.

As a section commander at PIT, I was in charge of four flights of pilots who were training other pilots to become instructors. Another great job—without much travel. PIT was the same school where I had been Top Graduate a few years before. I really enjoyed training others to be instructors, and, honestly, I was good at it. And I was at home most nights and weekends.

Soon after I departed ATC Stan/Eval and joined the PIT squadron, now-Major Bill Blackmon was scheduled to depart in June of 1981 to attend Air Command & Staff College at Maxwell

AFB in Montgomery, Alabama. Maxwell had an annual airshow in April of each year, where there was a large array of military and civilian aircraft on static display, as well as aerial performers. As Bill would be moving to Montgomery soon, he requested to be one of the pilots to fly a T-37 for static display at the airshow and, hopefully, have time to look around for a place for his family to rent for the next year.

Bill asked me if I'd be interested in going with him to the airshow and standing by the plane while it was on display to answer visitors' questions. That would provide Bill some time to look at the rental market around Montgomery. Since I was probably going to be attending ACSC the following year, and I thought the trip with Bill would be fun, I agreed.

We flew the Tweet to Maxwell AFB, and I stood by the plane during the sunny two-day airshow, answering questions about the T-37, Air Force pilots, and anything else the attendees asked me. By Sunday evening, Bill had found a nice house to rent, and I had a terrific sunburn on my entire face except for the triangular-shaped area at the top of my forehead that had been covered by my flight cap.

Needless to say, the flight back to Randolph with the helmet pressing on my sunburn made for a painful trip. But, in the long run, the trip also paid benefits for me since I ended up renting the same house that Bill did when we moved to Montgomery the following year. That saved me from having to find a place on my own.

The Air Force in my day was less formal and, I think, a more "fun" organization than it appears to be today. Computers were just beginning to appear at the squadron and lower levels and were usually overseen by trained NCOs—probably so overly confident officers didn't totally mess them up.

One example is the Air Force of my early days didn't take physical fitness tests seriously; at least in my experience, that would be an understatement. After surviving basic or officer training, any individual facing the annual physical fitness requirement simply needed to run a mile-and-a-half in about twelve minutes or so. No push-ups, no sit-ups, no weight lifting—just run that distance each year within the time limit. Hey, we were in the Air Force, not the Marine Corps!

As one advanced in rank and position, the fun factor took a steep decline, as junior officers could get away with the silliness that field grade officers couldn't. The higher I climbed in rank and positions of responsibility in the Air Force, the more I had to try to behave, as an example to my subordinates. But every once in a while, my innate sense of fun would surface, and I'd do something I considered funny—but perhaps not totally appropriate—for a major and section commander in a flying squadron.

Once I graduated from pilot training, received my wings, and reported to a tactical flying squadron, I never had to actually run the annual mile-and-a-half for time again. Yes, we were all *supposed* to do the annual run, but most people didn't. Instead, each year someone in the squadron would ask, "How fast did you run?" and every year, everyone would just make up a time (well below the max time allowed), and life would continue smoothly.

That's not to say I never exercised since I normally ran one-to-two miles a few times each week, but on my own and not timed. As a pilot, staying in good physical condition was necessary in order to endure the physical demands of flying, such as being able to pull 3-4+ Gs without blacking out if your heart wasn't strong enough to pump blood to your brain.

As it turned out, one year, someone decided that all of the personnel of the PIT squadron—my squadron—would actually run the 1.5 miles. *What?!* But, being good GIs, we all showed up on the designated morning on the outer perimeter road at Randolph for a timed squadron run. After a brief warm-up, the group was off and running, with Colonel Rogers leading the pack.

I started running and was doing just fine when temptation raised its head in the form of a base fire truck coming up from behind. As the fire truck approached, I stuck my thumb out as if I were hitchhiking, and, to my surprise, the truck actually stopped for me. Seeing this as an unexpected blessing, I jumped on the back of the fire truck, and off we went. As we passed the pack of sweating squadron mates, I, of course, had to yell, "See you at the finish line!"

I'm pretty sure I heard Col. Rogers yelling, "Jackson! Get off that truck and run!" But…

I was waiting at the finish line, eating a donut, when the rest of the squadron eventually finished. I can now reveal that Col. Rogers admitted it was the funniest thing he'd seen, and he didn't make me redo the run. It's good to have friends in high places.

In September of 1981, Karen and I welcomed our daughter Lori to our family with her birth at Wilford Hall Medical Center on Lackland AFB. We were obviously in heaven. As I have never smoked in my life, and I also can't stand the smell of cigars, I was not going to pass them out to celebrate the birth of our baby girl. Instead, I drove to the local Dunkin' Donuts and ordered fifteen dozen glazed donuts with pink icing. I picked them up the next morning and delivered them to all the places on base that I had frequented.

After teaching for two years, Karen took a year of maternity leave to stay at home with Lori. By mid-1982, life was good! We had a beautiful new daughter, and I had completed my MBA. We both loved living in San Antonio. I enjoyed working at Randolph AFB, flying the T-37, and being around our wonderful group of military and civilian friends. However, I had been promoted to major and selected for Air Command & Staff College, which meant I was being groomed for possible promotions and higher-level positions in the Air Force. It all looked fantastic. Too good to be true?

One of the added perks of being stationed at Randolph AFB was that Randolph was the location of Headquarters, Military Personnel Center (MPC), the place where <u>all</u> Air Force assignments were generated. In anticipation of my assignment after my ACSC graduation, I hoofed the two blocks from the Tweet squadron to MPC to talk with the resource managers in the

Air Training Command rated-pilot section about potential follow-up assignments.

Ever since I transferred from Wright-Patterson AFB and B-52s and arrived at Reese AFB in 1975, I worked under the assumption that I was now known as a "career trainer." As such, I would remain in Air Training Command for the rest of my career unless a better opportunity presented itself in another major command or headquarters. That was absolutely fine with me.

One of the career trainer assignment officers was a former T-37 instructor pilot whom I had known for years, and he was a good friend. Since I had excelled during my time in ATC as an officer, instructor, and headquarters staff member, and I had maintained a good reputation within ATC, I figured I would have an easy time selecting my ATC assignment after ACSC. I soon discovered that I had another thing coming.

After locating the ATC assignment section, I looked for my friend, whom we'll call "Captain ATC Assignment Guy" for his own protection. Certainly, my good buddy would help me get a fantastic ATC assignment after ACSC, right?

Things started out great. My friend was happy to see me, and I could immediately tell that he would be more than willing to help me get a good, career-enhancing assignment after my schooling. After all, I was a poster child for "How to Progress in a Pilot Career" and had held about every position within the UPT environment, from line instructor pilot to one of only five T-37 command flight examiners in the entire command. I had commanded flights through section levels and had performed well enough to, hopefully, have those above me sufficiently impressed to back me for positions as operations officer and then squadron commander following ACSC.

Seemingly, I was on my way to bigger and better things in ATC, and nothing could stop me now. Well, maybe there was *one* little thing ...

As my (nameless) friend pulled up my records on the personnel center computer, he was totally upbeat. When his smile quickly faded, I knew I was in trouble but had no idea why.

He turned away from the computer, faced me, and inquired, “Do you know you’re in the SAC Swap program?”

I froze. I knew that there was a program that allowed SAC personnel to take an assignment in another major command with the understanding it was a one-time-only assignment and then go back into SAC. “I didn’t think that program had anything to do with me,” I responded.

“Anyone in the SAC Swap program has a prefix at the beginning of their AFSC. Yours has it.”

The Air Force Specialty Code (AFSC) is used to identify specific jobs within the Air Force. For example, my AFSC as a T-37 instructor pilot was 1495D. In all of my time in Air Training Command and T-37s, I had never seen a prefix used before it. As it turned out, the Personnel Center computer system *did* have a prefix before my AFSC, and that meant I was a SAC Swap asset, and I would return to SAC following my year at ACSC. The fact that I had been in ATC for three different assignments at two bases, Reese and Randolph, meant nothing to the computer. My world had just turned completely upside down.

I did mention that this captain was a friend of mine, right? He could have delivered one of the usual personnel weenie speeches: “The assignment system is for your own good,” or “The needs of the Air Force come first,” or “Your ‘special skills’ (whatever those might be) would be of greater value in SAC.”

Instead, he saw the look of panic in my eyes and simply said, “Let’s go talk to the SAC pilot personnel guys.”

I followed him through the bowels of the large, winding MPC building to the side where the SAC assignment people worked. As it turned out, I wasn’t my captain’s only friend. He was also friends with one of the SAC pilot assignment officers.

My ATC captain, the SAC personnel officer (who will also remain nameless), and I sat down at the SAC pilot's desk as my captain explained my virtues: I was one of ATC's best instructor pilots; I had been a student squadron commander as a captain; I had been one of only five T-37 command flight examiners in ATC headquarters, and I was now a section commander at the ATC instructor pilot school.

He also explained that he and I were friends, and I had absolutely no idea I was in any kind of SAC Swap program, plus I had served at two different bases within ATC. And he mentioned that I was soon to leave for Air Command and Staff College, and ATC wanted to keep me within the command to return as a possible operations officer in a flying squadron. He pulled out all of the stops on my behalf. At this point, I was ready to kiss the guy!

"There's no way that I can undo a SAC Swap-identified pilot," the SAC officer told us. Then he paused briefly and added, "Let's make certain he's really in the SAC Swap program," which I thought was strange since my captain had just shown me the prefix on his computer. *Really* strange…

The three of us looked at his green computer screen, and the SAC officer said, "Yes, you're in the Swap program," because the prefix he highlighted with the blinking cursor on the green computer screen definitely was a swap prefix.

He then commented, "It's too bad that prefix is there because that's all that identifies you as a Swap pilot."

At that point, my buddy asked the SAC officer slowly and deliberately, "So, if someone were to hit the delete button, that SAC Swap identifier would be gone forever?"

The SAC officer smiled and said, "Yes."

At that point, my captain asked the SAC assignment captain, "Do you want to go get some coffee?" The SAC pilot's response: "Sure." They left me behind. Alone. With the computer. With the delete button.

I took a quick look around the office and saw no one, so……I pushed the delete button, and the prefix that the SAC officer had so conveniently highlighted disappeared. The SAC Swap prefix was gone. Now all I had to do was hope that I had correctly interpreted the two assignment guys' comments as they intended, and there wasn't a court martial in my future.

When the two captains returned from their coffee break, the SAC officer looked at the computer screen and said, "I guess we were worried about nothing. There isn't a SAC Swap prefix before Scoop's AFSC, after all. He belongs to ATC."

I can't remember ever being so relieved or so happy! (Well, at least not since I exited Vietnam.)

I was officially an ATC resource, and Captain ATC Assignment Guy was my new best friend. He and I returned to his office to discuss my ATC assignment possibilities after Air Command and Staff College.

The rest of the story is that I later learned that the SAC assignment officer had also flown O-2s as a forward air controller in Vietnam and had seen my FAC assignment on the computer screen as he brought up my records in his office. I think that, first, he was a good guy trying to help a friend and a friend of a friend. Second, he thought he and I had already paid our dues to the Air Force as FACs doing a dangerous job under lousy conditions in Vietnam.

There are moments that change your future. Sometimes you realize it, and sometimes you never know it happened. This is a case where I knew it happened, and I will always be grateful to those two gentlemen.

I had sworn after my assignment misfire, coming home from Vietnam to a B-52 assignment rather than my "promised" fighter assignment, that I would never let the Air Force assignment system "stick it to me" again.

First, Colonel Ullrich had gotten me out of SAC, and now, Captain Unnamed and a sympathetic SAC officer had kept me from going back to SAC. Thanks to good friends and good luck, I never did. I was to stay in Air Training Command for the rest of my Air Force career.

How long could I keep working the system? Stay tuned.

After two-and-a-half quick years, my time as an instructor pilot was over, and Karen, Lori, Barney Dog, and I packed up yet again. With a nine-month-old infant, this move was a little more logistically complex. Thank heavens, my mom and dad flew to Texas to help.

So, we formed a caravan of three cars as we drove away from our recently-sold house on Miller's Ridge Drive, with many fond memories of the years we'd spent in San Antonio and the great friendships we'd made. In the heat of the summer of 1982, we headed for our next adventure at Maxwell AFB in Montgomery, Alabama, where we knew there would be more good memories in the offing.

Mike's T-37 "Fini Flight" and the traditional hosing down. He flew the flight with his friend and commander, Lt. Col. Jack Rogers, and his dad had the privilege of turning on the fire hose.—We think he enjoyed it.

Chapter 6

First One School, Then Another

My year at Air Command and Staff College (ACSC) at Maxwell AFB in Montgomery, Alabama, was, in a word, academic. Not academic good or bad—just academic. Let's just say that sitting in a classroom or auditorium all day and doing homework at night for almost a year wasn't my idea of fun, and it was something I hadn't done since college. (Truthfully, I rarely did it in college.) Some of the students enjoyed ACSC, but I missed the excitement and freedom that I'd gotten used to during my twelve years of flying.

What I did *not* miss was the constant neck and back pain that was exacerbated by flying. In that respect, Vietnam never really left me. The constant pain and discomfort I felt was a daily reminder of my injuries from the rocket attack ten years earlier. Being a military pilot is what I was, and had been practically since my graduation from college, but I was beginning to question the price I was paying on a daily basis.

On the positive side, I had regular class hours and was home on time for dinner every night. And, given that we had our one-year-old daughter Lori to baby and play with, I was enjoying this new lifestyle. There was even an opportunity to occasionally venture out for pie or for dinner. I studied every night, but I'd gotten used to doing that while I earned my MBA last year just before Lori was born.

Maxwell AFB is a beautiful Spanish-styled base with a long history of aviation going back to the establishment of a flight school on the site by the Wright Brothers in 1910. The base was the headquarters of Air University and was largely dedicated to post-graduate professional military education for officers of ranks from lieutenants through colonels.

Air Command and Staff College students were almost exclusively majors, all of whom were unofficially being looked at for further promotions and responsibilities. Unlike my undergraduate education at Ohio University, there was little time for much other than studying, and one of the graduation requirements was writing a research paper which was similar to a post-graduate dissertation. There wasn't a lot of screwing around here. OK, there was *no* screwing around here, but it was still much better than the constant pressure I had experienced as an undergraduate pilot training student twelve years before.

We grew to like the city of Montgomery. We explored the historic sites in the downtown area and also took drives in the rural surroundings, where we quickly learned to identify kudzu. We'd never seen it before, and it seemed to cover everything along the highways. The most invasive species in our home area is honeysuckle, but it is no match for kudzu's power to entirely cover an abandoned farm building. Unbelievable.

We developed a habit of eating breakfast on Saturdays at the nearby Cracker Barrel, something else that was new to us—both eating breakfast out and the restaurant chain. The selection on the menu was far more limited than it is today, and we have a one-page paper menu to prove that framed and hanging on our kitchen wall.

We found the weather in Montgomery to be pretty close to ideal for us, except for the hot, muggy months of August and September. Our neighborhood had a walking/bike-riding path through its center, and Karen often placed Lori in a bike seat and rode the path after lunch. Then Lori would take a nice, long nap, enabling Karen to work on quilting and other sewing projects. So, she and I had switched roles from her spending days in a classroom to my doing the same.

After almost a year, in June of 1983, my ACSC course was completed, and it was time to graduate and move to our next operational assignment. The folks at the Military Personnel Center initially had me programmed to go to the Pentagon in some sort of staff job. I'd heard my ACSC classmates who had previously been assigned to the Pentagon talk about the long hours and long commutes, and I didn't want to do that at this point. Having a baby and being home regularly had spoiled me.

That involuntary trip to SAC had stuck with me. Therefore, as I'd done before, I hoped to have some control over my next assignment. Onto the phone I went to see if 1) I could find any friends in the personnel side of the house at the Military Personnel Center back at Randolph and 2) They still remembered "the old Scoop." I did, and they did.

I was warned, however, that neglecting to "fill the Pentagon square" on my resume would hurt my chances for further promotions, but I had grown to appreciate spending time with my wife and daughter on a regular basis, so I decided family trumped career. I also figured I had a good enough record at this point to make lieutenant colonel, and I'd worry about making colonel if I decided later that I wanted to stay in the military for more than twenty years.

Karen and I had grown to love Texas, and San Antonio in particular, so I finagled an assignment as a basic training squadron commander at Lackland AFB on the west side of San Antonio. (Our previous base, Randolph AFB, was east of town.) I eventually asked to have that assignment changed so I could become a squadron commander at the Officer Training School (OTS) at the Medina Annex, also at Lackland. OTS was the place where I got my start in the Air Force in November of 1968, leading to my being commissioned as a second lieutenant in February of 1969. Returning there was truly coming full circle.

So, a year after leaving Randolph AFB and San Antonio, I graduated from Air Command and Staff College and was off to Lackland AFB, the "Gateway to the Air Force," in—where else? … San Antonio! Once again, Karen, Lori, Barney Dog, and I were off to our next assignment.

This time we purchased a home on San Antonio's north side. It was the perfect location for driving anywhere in the city—shopping malls, eating establishments, the airport, and the Riverwalk. This was a spot where the Jacksons could be happy to stay forever. Maybe. Maybe not.

The military uses the term "career broadening" for pilots to say, "It's time to do something other than a flying job to show you *can* do something else." No airplanes in this assignment, but, as it turned out, there were some things I enjoyed just as much, if not more.

Heading to a position as a squadron commander at OTS excited me for several reasons: I'd be back in San Antonio; my neck and back would get a continued reprieve from pulling Gs and being bounced around in an airplane; there would be good hours for family time; and I would have an additional opportunity to apply my bachelor's degree in education. As an officer trainee (OT) when I attended OTS in 1968, I thought it seemed like it'd be a good place to work. As Karen had discovered in her short tenure at Texas Tech University, the students were highly motivated, and there were no discipline problems. I never for a moment thought I'd ever end up back there if I finished pilot training. Life was good, and I was happy. So was my family.

First, before I could become a full-fledged squadron commander, I had to "carry" a flight of twenty trainees through the twelve-week course, leading to their commissioning as second lieutenants. Of course, things are seldom as easy as they sound. Before I was allowed to carry a flight of students, I had to become certified as an ATC officer trainee instructor. Yet another round of certifications, this time a three-week program involving teaching techniques, public speaking, writing, Air Force history and traditions, and—by far the worst for me—drill and ceremonies, as in *teaching trainees to march*. Yikes!

Marching? Don't they know that *no one* marches in the Air Force? That is actually *almost* true. The Air Force rarely marches except in training. The only time I'd marched since I was a trainee at OTS was when we had a one-star general as wing commander at Reese AFB who thought (unwisely) that he could distinguish his wing above the rest of the UPT bases by having a graduation parade. Even more of a mistake was including the staff of pilot officers in the parade. It was a total disaster! Crooked lines, the majority out of step… It was the first and *last* time anyone *ever* heard of a pilot-training graduation parade. Anywhere.

Having me teach anyone to march was akin to asking me to teach someone to fly without wings. Hopeless. *I'll just fake my way through it,* I thought. Fortunately, the unlucky captain running the check-out program saw the big picture. I was probably the worst drill and ceremonies instructor he—maybe all of OTS— had ever seen. "I know you'll have to teach marching to just one class of students," he concluded, so he let me sneak through that part of the evaluation. I will say that my performance in the remainder of the instructor checkout was excellent, as my college degree in education and more than seven years of experience as an instructor pilot gave me a huge advantage.

One thing I always appreciated, both as an instructor pilot and then as an OTS instructor, was my students were well-screened and some of the top graduates from their colleges. To rank in the top tier of the competitive application process and be offered a position in the OTS training program, a college graduate not only had to do well in his/her university performance but also on the Air Force Officer Qualifying Test. Then, students who wanted to be selected to continue on to UPT had to finish at the very top of their commissioning source (OTS, the Air Force Academy, or the AF Reserve Officer Training Corps). Those factors made my job much, much easier and a lot more fun.

A few students at OTS were previously enlisted personnel, referred to as "prior service." These men and women also had earned a bachelor's degree and were selected in competition with other active-duty enlisted personnel for a very limited number of positions available to prior service candidates. Again, cream of the crop.

The Air Force is wonderful in the way it welcomes new people to an organization. In past moves to new units, I'd always received a letter from a sponsor who would welcome Karen and me to the new unit and be available to show us around, answer questions, make introductions, take us out to dinner, and involve us in the social whirl of the new unit, base, and town. At the time I arrived at OTS, those officers had a different tradition. They offered to come to your house to build a new patio cover, install an in-ground irrigation system, or build a storage shed.

This was a welcomed offer for new arrivals since nearly all of us lived in new or existing homes that could use at least one of those items. All a recipient was required to do was pay for materials, help in the construction, and provide food and drinks. (What a deal!) Then, if willing, to participate in future projects. I accepted their offer to build a backyard storage shed. This shared construction is how I learned the procedures to install a lawn irrigation system, a skill that would come in handy more than once.

By September of 1983, my training was completed, and I picked up a flight (class) of 20 new Air Force trainees assigned just to me. They were a mix of young, recent college graduates and a few older prior service ex-sergeants. (They held no rank as officer trainees.) I liked them all, and they ended up being competitive for top flight honors within our squadron and the group.

The officer training program begins with a lot of discipline and a "spit and polished" approach. There is little freedom initially for the new trainees. As a flight commander, I oversaw and instructed the academic side of officer training. The course lasted twelve weeks. There was always an upper class six weeks ahead of the new lower class. OTS expected the upper class to have learned enough by the sixth-week point to assume the role of instructors for the lower class.

Luckily for me, one of the duties of the upper class was to teach the lower-class members to march. I was saved! In their role as leaders, they were also the people who, in general, ran OTS in the after-duty hours after the commissioned staff went home. The system worked well, as it had for decades, and I was continually impressed by the level of training and professionalism that was instilled in the OTs after only the first six weeks.

The program eased up progressively as the trainees progressed through the twelve weeks and were continually given more responsibility. By week three or so, if they received few demerits, they were permitted to leave the base on weekends and enjoy all that San Antonio had to offer. I doubt there are many people who have been to either basic training or officer training at Lackland who don't have a photo of the Riverwalk or Alamo that was taken on a free weekend.

My flight was still in training over Thanksgiving weekend, as I had been many years prior when I was a trainee. As it was a day off of training, Karen and I invited them to our house for an afternoon and a chance to relax away from the OTS environment. I think they all enjoyed being reminded that there *will* be life after OTS, and Karen and I enjoyed getting to socialize with them in a non-classroom environment. Of course, the next day, it was back to the old grind for them, but they were far enough along in their training to see the light at the end of the tunnel.

While the trainees were at our house, I couldn't pass up the opportunity to show them one of Karen's superpowers: remembering names. She claims it was an after-effect of the pressure she felt as a pledge for her sorority, Alpha Delta Pi, at Ohio University, when she had to identify every one of the seventy active members by name and hometown in a "line-up." I'm one of those people who can remember faces easily, even years later, but

I can't remember a name five seconds after I've heard it. Karen is the opposite; she remembers names instantly but not faces.

Karen had studied a class roster the day before, but she hadn't seen photos of the OTs. I asked the trainees to sit on our living room floor in a circle and tell Karen their first and last names. She then went around the circle again and flawlessly identified all twenty of them by their first and last names. They were impressed, as I'm always impressed by her memory—except if I've done something she didn't like.

My twelve weeks as a flight commander passed quickly and pleasantly. I was due to take over a squadron now that I'd trained one flight of trainees, and I'd been given a squadron commander's office before my flight of soon-to-be lieutenants graduated. Before they left for their assignments, two artists in the flight painted a beautiful eagle with an F-15 in flight in front of it on my new office wall. I absolutely loved that token of their appreciation and made certain nothing was ever placed over or in front of the painting. I've been told the painting remained on the wall for many years after I left until OTS was relocated from Lackland AFB to Maxwell.

After three months, my first and only class of Jackson-trained second lieutenants were graduated, commissioned, and off to do wonderful things in the Air Force, and I moved on to the squadron commander position. Having been a flight and section commander in the high-pressure, high-tempo, and sometimes dangerous UPT and PIT squadrons, I found this job to be fun and relatively easy.

The OTS commander, Colonel James Weaver, was a friend of mine from my days flying T-37s at Reese AFB and Randolph AFB. I should also mention that our OTS Vice Commander was Colonel James Fleming, who had been awarded the Medal of Honor for his courageous actions in Vietnam. Two gentlemen and fine leaders. They were wonderful examples of professional officers for both the OTS staff and the trainees.

The trainees were subject to many rules and regulations, more than a few of which seemed petty and unnecessary at first glance. Actually, there was a valid purpose. The rules were instituted to teach attention to detail. I admit I didn't understand the need for most of these rules myself when I was a trainee, but I came to understand the lessons they instilled in our future officers once I got out of training and into "the real Air Force."

In reality, those seemingly petty rules taught the need to be constantly on alert for details. Missing—or overlooking—a small but important detail could cause harm to themselves, to others, or to equipment in their post-training jobs. We were teaching habit patterns, careful attention, and following checklists, whether they were written or memorized. I find, even today, I'm focused on small details that others usually don't notice or think of.

The result of not following a rule at OTS was the issuance of demerits. OTs hated demerits, of course, and for a good reason. Consequences ranged from lower rankings by their instructors on performance evaluations to—*The Biggie*—getting "gigged in" for the weekend, as in being restricted to their room or to the OTS campus. It was the civilian equivalent of being grounded. It meant not going into San Antonio, or it could mean extra duties.

For example, dust on the floor of their room might earn five demerits. Failure to make the bed properly: five demerits. Moving while at attention: ten demerits. Failure to respond to a question or order immediately: ten demerits. Failure to salute a superior: fifteen demerits. The only way to erase demerits was by earning merits. Merits offset demerits but were hard to earn, and the OTs would do many things to earn merits.

True to form, I was unable to remain the complete officer role model at all times. For example, Colonel Weaver stopped me one day and inquired, "What's this I hear about a donut machine?"

Flights of OTs sold donuts each morning from a table at the bottom of the stairway leading to the floor where my office was located. I *love* donuts. That meant I had to walk all the way down the stairs, buy a donut, and then walk all the way back up the stairs to my office. Rough life, huh?

Some of our OTs were engineers, and I casually mentioned to a couple of them, in jest, "Build a donut-retrieving machine so I can get my morning donuts without walking down the stairs. I want to be able to hook a donut and pull it upstairs. Oh, and by the way, there *might* be merits involved if successful." That was all it took to set off the "Great Donut Snagging Design Competition."

The next morning, and several mornings after that, there would be three or four OTs waiting for me at my office to try out their donut snaggers and hopefully gain some prized merits. Many of these creations were actually pretty ingenious, especially considering there weren't many resources on the campus from which to make the "snaggers." A crowd of OTs and staff started to gather every morning to watch me try out a new donut-grabbing device.

Unfortunately, Colonel Weaver got wind of the snagging competition and didn't see it in the same creative light that I had, although he *was* laughing when he told me, "Stop." That was the end of the donut competition, but I made it right with Colonel Weaver by taking him a few donuts. All was forgiven.

I did give out a few demerits, mainly if the OT had country western music on his or her radio when I turned on the radio if I inspected their room. I wrote it up as "trash on the radio." I didn't inspect many rooms, so there weren't many demerits, but the word spread. Consequently, there weren't many radios at the Medina Annex set to play country western music.

The highlight of our OTS assignment was the birth of our second daughter, Katie, in 1984. Once again, the birth was at Wilford Hall Medical Center. Quite a coincidence for a career military family: two daughters born at the same hospital during two different assignments. Texas has been very good to our family, and Katie has added much joy to our family ever since. I was enjoying the normal hours and being home every night that this non-flying job was allowing.

But my two years at OTS were about to come to an end when I was promoted to lieutenant colonel. I was named a group commander over four squadrons at OTS, and I was told to look for another position commensurate with my new rank. Once again, I wasn't going to leave this entirely up to MPC. I became actively involved in the selection.

At this point, Karen and I were not only considering *where* my Air Force career would lead me next but *when* and *where* the best place would be to retire. We knew we both loved living in Texas, and we'd made so many friends there that it would be hard to leave. After spending ten years in the South, it felt more like home than our native Ohio. After all, this was my fifth, Karen's fourth, and Lori's second assignment in Texas.

But neither of our girls had spent more than a few days a year with their grandparents and even less time with aunts, uncles, and cousins. We'd been moving them from one base to another every year or two and wondered if we were depriving them of the experience of establishing roots near our extended family. We were definitely happy living in Texas. Could we be happy—or even happier—around friends and relatives back in Ohio? Only one way to find out. I needed to get an assignment to Ohio.

Not as easy as it sounds, though. I was an Air Training Command (ATC) resource, and, as such, I needed to find a job in Ohio that was in ATC, but the only active-duty base in Ohio is Wright-Patterson AFB near Dayton. It is a logistics and engineering center. *Is there anything I would be qualified to do there?* I wondered. A definite plus would be its proximity to Tipp City, my hometown. So, I focused on opportunities at Wright-Patt.

You may recall that ten years earlier, I did everything possible to get away from there, but that was because of my job as a B-52 pilot. I had no ill feelings about the base, just the B-52 assignment. The SAC wing had since closed down, and there are relatively few pilots at WPAFB. There is a 12,600-foot runway, but, at times, there are no operational active-duty flying units.

As luck would have it, there were only two positions at Wright-Patt that I was qualified to fill as a lieutenant colonel, a pilot, and an ATC resource: the commander of a recruiting squadron and the professor of aerospace studies at Wright State University in Fairborn, right outside the gates. Unfortunately, neither of those positions was scheduled to be filled for at least another year. By that time, I'd have been reassigned someplace else for three years or until I decided to retire.

As I'm sure you recall by now, I vowed, after my less-than-desired B-52 assignment, never to let the Air Force system choose another assignment for me without my "help." So, where to start this time? I didn't have to wonder very long, as my Vietnam boss and the person who earlier helped me out of SAC, now full colonel Dale Ullrich, was the vice commander of USAF Recruiting Service. Pleasant coincidence or divine providence? Who am I to say?

I made a quick drive across San Antonio from Lackland AFB to Randolph AFB, the headquarters of USAF Recruiting Service. Soon, I was in Col. Ullrich's office, telling him of my assignment plight.

He was sympathetic. "The Wright-Patt recruiting squadron commander's position won't be coming open for reassignment for another year. But," he continued, "the position in Syracuse, New York, needs to be filled this coming summer. If you take that job and do a good job increasing their numbers and the squadron's position in Recruiting Service standings, I can have you transferred to Wright-Patt the following year."

After living in the South for most of the last fifteen years, I wasn't crazy about moving to Syracuse—especially after I did some research. Syracuse averages 120 inches of snow each year, but this sounded like the only way I could possibly get to Ohio. I thanked Dale for his advice and said I'd give him an answer after I had a serious talk with Karen about a move north.

Neither Karen nor I had seen snow in almost ten years, but the chance to get to Ohio—if I did well in New York—was a powerful incentive to take the Syracuse position. So, in June of 1985, we were off to New York for a job which I knew absolutely nothing about. All of us had a lump in our throats and, possibly, a tear or two in our eyes as Karen, Lori, Katie, and I drove away from our home on Peppermill Run in San Antonio. I was leaving a great job, our nice house, good friends, wonderful neighbors, and a good climate and heading toward the cold North.

In case you're keeping count, Lori was three years old and moving for the third time, and Katie was ten months old and moving for the first time. I'd lost track of how many jobs Karen had been forced to leave. I hoped I wasn't making a mistake.

My office as a squadron commander at OTS, my first real desk job

Mike in front of the beautiful mural two of his OTS students painted for his office wall (He was honestly happier than he looked.)

Chapter 7

Raise Your Right Hand and Repeat After Me

In June of 1985, Mayflower Van Lines packed up our household goods in San Antonio, and Karen and I moved our family 1,800 miles from our favorite house, our favorite neighborhood, our favorite state, and one of my favorite jobs to DeWitt, New York. This was quite a change—from San Antonio, the city of festivals, multiple military bases, warm climate, and sunshine, to Syracuse, New York, a city of overcast skies, a small Air National Guard (ANG) base, a cold climate, and snow. Lots and lots of snow. And, of course, to the totally unfamiliar work environment of the Air Force Recruiting Service.

Our first task had been to find a house for our family to rent, since I hoped we wouldn't be there for more than one year. This was not easy. Karen and I flew to Syracuse in April, after dropping our daughters off at my parents' house in Ohio, on a three-day house-hunting trip. Have I mentioned that it snows a LOT in upstate New York? Given that fact, our main objective was to find a house with a garage. Besides my not wanting to uncover my car every morning before I headed to work, Karen certainly didn't want to fight through ice and snow to load two toddlers into car seats.

There were houses with attached garages everywhere we looked. Just none of them were for rent. We quickly realized that

Syracuse was not the hot market for home sales or rentals that San Antonio was. It was an economically depressed region. We were surprised to learn how few houses were for rent and even fewer with garages, and a garage was a non-negotiable item for us.

Adjusting to all of the changes was difficult for both Karen and for me. We were leaving the perfect neighborhood in San Antonio, and we weren't willing to live on a busy thoroughfare. After fruitless hours scouring multiple areas north and east of Syracuse, we knew we'd have to lower our expectations.

We found just one house that met our needs, and it definitely wasn't our dream property. It was situated in a nice neighborhood, with a playground close by and convenient shopping. Driving up, we saw that the landscaping was a bit overgrown, but that was manageable. The inside was a different story. The current renters included a heavy smoker, and, at some point, they or previous renters had at least one dog who evidently rarely made it all the way outside when needed. But the landlord *assured* us the house would be *thoroughly* cleaned before we arrived. We took him at his word and signed the lease. We didn't feel we had much choice.

Apparently, our landlord's standards of cleanliness and ours were worlds apart. We arrived to find that absolutely nothing had been done when the former tenants vacated. However, we were stuck. What else could we do other than clean everything ourselves? Fortunately, my parents had accompanied us on the Tipp City to Syracuse part of the journey. My mother was a fanatic about keeping a house clean, so she was a HUGE help in making the place livable, and Dad was a big help, too.

While Karen and Mom washed every wall, every window, and every door in every room, Dad helped me tame the lawn and shrubbery. I rented the most powerful carpet-cleaning equipment I could find, and I cleaned the carpet not once, not twice, but THREE times. Yep, *three* times! After all, Katie was a crawler at that point, so I cleaned until the rinse water cleared. What I couldn't clean—the hall and master bedroom carpet with holes and the unpleasantly-scented sun porch carpet—the landlord reluctantly agreed to replace.

Within a couple of months, Karen also convinced the landlord to allow her to replace the 1960s-era, full-of-tears wallpaper in the kitchen and both bathrooms—at our cost and with her labor, of course. When our next-door-neighbor Kathy—a friendly, fellow stay-at-home mom—told Karen she thought she was crazy to work so hard to improve a place we didn't own, Karen responded, "This is our home. Our only home right now. I want to make it as pleasant as I can." And she did.

So, the house quickly improved. Not perfect, but we'd be leaving in a year, we hoped. We also contracted with a snowplowing service to clear our driveway whenever it snowed more than two inches. We saw a lot of him that winter.

To say I felt outside my comfort zone when I first reported to work with the 3513th USAF Recruiting Squadron (USAFRSq) would be a vast understatement. In all of my previous assignments, I was surrounded by officers who were pretty much like me—pilots or educators in a different, but familiar, career field. Suddenly, I found myself an Air Force minority. As the commander of a recruiting squadron, I was outnumbered by enlisted recruiters doing things and speaking a language totally foreign to me. What the heck was an ASVAB, an NPS, or a MEPS? Monthly goals for everything from basic trainees to doctors and nurses?

To my newbie's eyes, recruiters struck me as the used car salespeople of the military, selling a four-year enlistment as an airman no-striper for a minimum wage, a uniform, and a chance to get killed in a war. That view changed quickly. My opinion of their job and their worth to the military skyrocketed as I became familiar with their many challenges.

There was an additional factor I hadn't experienced before. As opposed to the endless training and evaluations I'd received while a pilot and as an OTS instructor, I received little training to prepare me to command a squadron where I had so little idea what was happening. There was a *huge* learning curve. Getting used to being a recruiting squadron commander took some time, but thank heavens for the senior NCOs and staff in our squadron headquarters. They were the backbone of the unit. I'm assuming they had seen a series of "clueless" squadron commanders come and go over the years, and I was just one more.

As the commander of one of the thirty-five Air Force recruiting squadrons in the United States, I oversaw forty-four recruiting offices across New York and Pennsylvania. There were a total of approximately fifty recruiters assigned to those offices, and they operated pretty much on their own to represent the Air Force and meet the goals within their assigned areas. In addition to myself, they were supervised by the squadron's chief master sergeant (titled the squadron superintendent) and flight supervisors, who were senior NCOs. Each flight supervisor was responsible for approximately five to eight individual offices within a zone. As you can imagine, I relied heavily on the NCOs.

The recruiting branch of the Air Force does not operate quite like the flying squadrons I had experienced, where personnel rotated every three-to-four years. Recruiters manning the local offices in small and large towns across our region could remain in their location for multiple years if they were sharp and met their quotas. That enabled them to settle in with their families and get to know the personnel in the high school and other locals who were valuable in gaining access to possible recruits.

If the recruiters performed their jobs as intended, things would run smoothly. However, some problems arose due to the fact that recruiters were semi-autonomous in their day-to-day activities. Many were alone in their office and located many miles from their supervisors or other Air Force personnel. As an example, *my* nearest supervisor, my group commander, was a colonel based 310 miles away in Boston at Hanscom AFB. Not only were my recruiters geographically separated from their supervisors, but so was I.

When I arrived, the squadron's headquarters was located in an office building in downtown Syracuse. It was a central location, but there were two issues that required an immediate remedy. The structure was a rather small, multi-storied older building which we shared with other tenants. Our portion of the building was much too small for the staff to have sufficient room.

The second problem: everyone on our staff had to pay for parking in a commercial lot behind the building, and it wasn't cheap. If I remember correctly, it was about $75 a month to park in a blacktopped lot, and that was the cost nearly forty years ago. Some of our junior enlisted troops couldn't afford that amount, so they walked a distance from where they could find free parking space. Therefore, they usually had to walk for blocks from their car to their office, often in the rain, cold, or snow.

Thanks to the efforts of our staff, and some assistance from group headquarters, we found a solution to both problems. We moved from downtown Syracuse to a large two-story "bunker" on Hancock Air National Guard Base outside Syracuse, known as the SAGE building for Simulated Air Ground Environment. The largely-uninhabited building was solid concrete with no windows and two-foot-thick reinforced walls. Other than a small contingent of security personnel who protected the building and a few possibly government communications workers, we had the building to ourselves.

Since our new headquarters had security forces on duty, we had to show identification to be allowed through the gates, even though we were about the only people in the building. But there were many perks. There was plenty of free parking along the fence surrounding the facility, and we were able to park all of our government cars inside the gate. Initial problems solved!

We were told the structure served as an alternate White House during the height of the Cold War. It was a relic which intrigued a history education major like me, and so, of course, it became a tempting place to explore.

The two-story building was divided into two connecting sides, and much of it was now empty. One side had housed two 1950s-era computers, each filling one entire floor, and we were told that when in use, they generated sufficient heat to be piped throughout the building—no minor feat in the cold winters of upstate New York. The computers were removed years earlier, but the holes in the floors and ceilings for the connecting wires and other equipment were still there.

There was a large two-story amphitheater area with a semi-circular, continuous desktop and an extremely large wall screen. At one time, this was where military generals, and possibly the

President, would have been briefed in the case of a national defense emergency.

The other side of the building was filled with empty offices, and we took up residence on the second floor. There were so many offices available that I let everyone pick their own area. I, of course, picked the largest office, which supposedly had belonged to a two-star general. It had real wood paneling, a hidden door to a bedroom, and a bathroom with a shower. Pretty cool.

That side also had a large room, about 2,000 square feet in size, that was filled with floor-to-ceiling air filters, which were still being changed regularly. We were told by the security guards that the filters were installed when the building was constructed to filter out any nuclear or biological contaminates in the event of a war. We were also told that the building had its own water supply, although I never saw any evidence of it.

Nothing on the office side of the building had been used for years, so every part of it, except for the air filter room, needed to be cleaned. There was a lot of work to do. Every person pitched in for the rehab.

Since we were on Air Force property, I got the Air Force to supply us with new ceiling tiles for the staff to use if they really wanted to improve their areas. Karen helped me with my office. Our first project was the wood paneling. No small task since it covered every wall. We used two cans of Liquid Gold to restore a sheen to the dried wood. Then we washed the floors in the office and bedroom area and sanitized the bathroom. The bedroom became a storage closet. I never used the shower (but I did use the bathroom).

With the roomy facilities and free parking, everyone was happy. Even Lori and Katie benefitted from my new office because they'd sometimes accompany me to the office on weekends, and we always had to ride up and down many times in the large freight elevator with the big, heavy wooden door. It was a thrill for them. Other than when we moved our furniture and supplies into the building, squadron personnel used a mundane stairway. Boring, perhaps, but its usage avoided my writing safety reports.

In addition to the main task of supervising recruiting offices, all recruiting squadron commanders were part of the management team of a MEPS, the Military Entrance Processing Station, within their squadron's geographical area. This was an administrative and medical processing center where all potential enlistees, officer and enlisted, for all branches of the military went to be tested in order to get their academic qualification and pre-enlistment physical.

In the case of my 3513th USAFRSq, there were two MEPS, one in Syracuse and one in Buffalo. Commanders of the services that used the MEPS—the Air Force, Navy, Army, and Coast Guard—met at least once every three months to discuss various issues concerning the operation of the MEPS and the processing of their service's applicants.

Another duty of a recruiting squadron commander was to give public presentations about recruiting procedures, and military life in general, to anyone and everyone who would listen. Upstate New York didn't possess the military presence that you find in places like San Antonio, San Diego, Washington, D.C., or even Dayton, Ohio. Because there were no large Air Force bases, most people had no first-hand knowledge of military life. Therefore, my task was to help spread the word.

As the mouthpiece of the Air Force within my two-state squadron, I was regularly called upon for public speaking appearances to audiences ranging from school classes, local and state politicians, and media outlets. At first, I was reluctant to do these presentations—largely because I knew so little about the day-to-day job of recruiters and, therefore, what I was supposed to talk about. But, as I became more knowledgeable of the world of recruiting and felt more at ease speaking to groups, large and small, I started to enjoy that part of the job. As it turned out, this media experience would greatly benefit me in later years as I moved into post-Air Force positions.

I gave presentations throughout our squadron's boundaries and, as such, spent a lot of time driving up and down the New York Thruway and along the Great Lakes shore. I am always early to anything I'm scheduled to attend, sometimes *very* early. I was once due to give a talk to a group after a dinner meeting, and I arrived well in advance to make certain I could find the venue near the Buffalo area. With nothing to do until dinner time, I drove around the city and the dock area.

I grew curious about the large ships docked along the river since we have nothing larger than a canoe in Tipp City. Again, being quite early and also curious about what a large ship looks like on the inside, I decided to try to take a closer look at one that was offloading German cars. I was driving a government motor pool car that was blue, plain, and sporting U.S. government license plates, and I was wearing a uniform with a plain blue zippered jacket with lieutenant colonel clusters on the shoulders. If I didn't put on my Air Force flight cap, I thought I might look like some sort of Navy or Coast Guard officer. It was worth a try.

Figuring the best approach was the direct *I'm official, and I'm not afraid of anything*, I pulled my nondescript government car sharply into a no-parking area right next to the ramp of the ship, and I walked up the ramp with all the authority I could muster.

When a crew member looked as if he were going to ask me what I was doing, I turned and asked, "Are these cars coming off this ship?"

Duh, of course! They obviously were, but that crew member was so taken aback by my seeming authority that all he could say was, "Yes, they are." With that, I strode confidently (outwardly, anyway) onto the ship while he stared at the idiot who had just asked the world's dumbest question.

Once on the ship, I wandered all around, from one deck to another, checking out the galley, the sleeping quarters, the cargo hold—all while the crewmembers I passed probably wondered, *What the heck is this guy in some sort of uniform doing on our ship?*

As I approached the ramp to disembark, a different crewmember approached me, undoubtedly to ask me what in the world I was doing on his ship. But, with all the outward confidence I could fake, I simply announced, "I think it's all OK," as I walked briskly past him and got into my car. As I drove away, I could see a cluster of crewmembers staring in a confused manner at the departing car, not realizing they had just been visited by a curious Air Force pilot.

In general, I came to enjoy serving with the recruiting service. Successful recruitment was, of course, the key to manning a productive, sufficiently large military force. The one MAJOR task I didn't enjoy, however, was the fact that I, as the squadron commander, was the person ultimately held responsible for the squadron's recruiting production and for its discipline.

On the production side, each of the thirty-five recruiting squadrons across the nation was tasked with a goal or "quota" of people to recruit each month. All types of enlisted and officer positions were included in our goals. That meant all career fields, from enlisted cooks, mechanics, administrative, and all other enlisted fields in the Air Force to potential pilots, navigators, lawyers, engineers, nurses, doctors, and all other officer career fields.

Each month, it was a struggle for the entire squadron to meet its goal. My job was to make certain that each recruiter was putting in the maximum effort to meet his or her own goals so the squadron could meet its goal and the Air Force, in turn, could meet its goal of the enlistments necessary to carry on its mission.

As is true with all groups of people in any career field anywhere, problems could arise. Despite the high standards an individual met to become a recruiter, there were always individuals who couldn't or wouldn't do their job by exerting the effort needed to effectively meet their responsibilities. Many of their issues were corrected through additional training from their flight supervisor or the squadron superintendent. The more serious issues ended up on my desk. Some situations, unfortunately, resulted in my taking administrative or judicial action, which could result in a recruiter losing his/her job and returning to the "regular" Air Force, a loss of rank, or an eventual court-martial. As much as I dreaded the necessity for corrective action, it had to be done in order for the unit to function effectively.

There was a lot of competition between the branches of the military to meet their recruiting goals. This translated to competition among the recruiters in the local offices of the different branches. It was common to have offices of more than one branch located adjacent to each other in a strip mall, for example. The different recruiters got to know each other well. Although the competition was usually cordial, it could become a bit devious at times.

In one instance, I was made aware of our potential enlistees consistently being diverted and recruited by a different branch. We couldn't figure out how the other military branch's recruiters knew the names, addresses, and phone numbers of our potential recruits in order to "poach" them away from us by making pie-in-the-sky promises before they enlisted into the Air Force.

After a little unofficial but highly effective detective work, we discovered that the recruiter of the service branch in the office next to our Air Force office was waiting until our office was empty and then climbing up through the ceiling panels, making his way across the ceiling, and dropping into the Air Force office. Once in our office, this other recruiter rifled through the Air Force recruiter's files, copied the names and contact information for our potential enlistees, and then returned to his office by the same route. He then called the person with all kinds of offers that were impossible to deliver.

For example, the rogue recruiter might call some of our potential recruits and say he could guarantee some amazing jobs if they joined them and not us. The guarantees were never honored, so we lost potential airmen, and the poached recruits lost much more.

Needless to say, I phoned the opposing branch's unit commander and gave him the option of disciplining the rogue recruiter—a possible court-martial for breaking and entering was on the table—or I'd call the police and/or the next person in the chain of command above that commander. He took care of it.

After that was remedied and the crooked recruiter removed (to who knows where), we then attached chicken wire above the drop ceiling, and all of my recruiters, squadron-wide, were told to lock their files whenever the office was empty.

Despite that one breach of etiquette, the branches and recruiters during my tenure usually got along well, since they faced the same challenges. If a potential recruit could not meet the standards of the Air Force, my recruiter was free to recommend the applicant to another branch, make any introductions, and pass on our test results or other government information. The various military branch commanders were friendly with each other and would try to meet for lunch after a MEPS meeting, for example, to compare notes and commiserate. "Cooperate and graduate" applied to another situation, I guess.

By the time I put in the year that we had agreed to, my squadron had moved up from the mid-30s in national standing to number 12. I had done what was requested of me with the help of the entire squadron. A promise was kept this time, and Recruiting Service Headquarters transferred me to the 3552nd USAF Recruiting Squadron at Wright-Patterson AFB.

Karen had made four friends in the neighborhood who were all stay-at-home mothers of preschoolers. They met monthly at each other's houses for coffee, allowing the children to play while the mothers enjoyed some adult conversation. In addition, Lori had become quite the proficient snow sledder on the slope at the Jewish Community Center across the street, and she liked the preschool she attended three mornings a week.

Thus, Karen was not eager to leave so soon, but I'd tired of the long, snowy drives up and down the Thruway, and we knew there would be advantages of living near family. In June of 1986, after eleven years away, Karen and I said good-bye to New York and headed back to our home state of Ohio.

- - - - - - - - - - - -

The Jackson family was much more eager to make the move back to Ohio than we had been for the move from warm San Antonio to the cold and snow of upstate New York. It would be wonderful to be near my family, who lived only fifteen miles away from WPAFB, and Karen's family lived relatively near in the Cleveland area. Lori and Katie could meet and get to know their many relatives in both locations.

Since the move to Ohio was originally sought to help us decide if we'd retire there or return to Texas, we decided to take our time before buying a house where we didn't know if we would reside for a couple of years or for many years. Instead, we moved into a two-story duplex on the base while we explored real estate possibilities in Tipp City. We expected to live on base for a relatively short time.

From the first day of our arrival, we enjoyed living on base. Karen and I lived in base housing only once before—during my first assignment after I completed pilot training—in a small duplex on Shaw AFB in South Carolina before I left for Vietnam. We'd loved the convenience and the comradery we'd found at Shaw, so we were hopeful for a similar experience. We weren't disappointed.

On the very first morning, as the moving van arrived to unload, neighbors came by to welcome us. One resident came with a clipboard and pencil. After introducing herself and pointing to her residence across the street, she asked us how many children we had and then recorded their names and ages. Another neighbor offered to take our children to her house to play with hers while we dealt with the movers. They couldn't have been nicer.

I guess it's just natural for Air Force families to waste no time forming friendships in new surroundings. It wasn't a case of nosiness; it was a means to welcome the military member, the spouse, and the children and help make their move to a new home as pleasant as possible. They'd all been in the same situation, probably multiple times. By the end of that first day, we knew we'd made the right decision to live on base.

Also, the location couldn't have been more convenient. We were in close proximity to the BX and commissary. The medical center was nearby, and there was a preschool on base. Lori and Katie had lots of kids their ages to play with on the large playground right beyond our patio door. All Lori had to do was look outside and say, "Mom, Sarah's out on the playground. Can I go out?" We weren't used to everything being so readily available and in a secure setting where you could keep an eye on a five-year-old through the back door.

Since my office was also close by, I got into the habit of coming home for dinner, spending some family time in the early evening, and then returning to the office about five blocks away to finish any paperwork later at night, if necessary. A person could get used to this lifestyle!

The transition from the recruiting squadron in New York to the squadron in Ohio was relatively easy, since I had survived the growing pains of learning about the recruiting world and come out of them pretty well-informed. I'd gone to New York not knowing a thing about commanding a recruiting squadron to now being a seasoned veteran.

The 3552nd included portions of four states within its boundaries—Ohio, Indiana, West Virginia, and Kentucky. Having grown up there, I was familiar with much of the area, including Athens, home of our alma mater, Ohio University.

My new squadron was in good shape when I arrived. The officers, the superintendent, flight supervisors, and recruiters were all performing their duties well, with only a few inevitable exceptions. The squadron had an excellent record for making its recruiting goals and even overproducing. (It definitely made a difference to have a significant Air Force presence in the area.) There were no problems with the facilities, and parking was free and abundant, of course. Most of the time, I felt as if the squadron could operate without me. *Most* of the time.

Of course, since there were human beings assigned to the squadron, there would be occasional problems, but I'd learned how unpleasant situations were handled in another squadron, so I had an advantage in this area, too. In Recruiting Service, there were few squadron commanders who had commanded more than one squadron, so I was in a pretty good position experience-wise.

As with my previous squadron, this unit had about forty recruiting offices and two MEPS spread across our four-state area. As I had done in New York, I tried to visit each office once a quarter, so I was gone a lot of nights, visiting three or four offices on each trip. At an earlier point in my life, I enjoyed driving, but not any longer. I grew tired of the repeated driving for office visits, speaking engagements, media interviews, the two MEPS, and various trips to show educators or other "influencers" Air Force facilities.

A rare benefit of traversing the highways and byways of Ohio was the opportunity to stop at Bob Evans restaurants. Hey, I had to find something to look forward to as I accumulated miles on my government vehicle's odometer. Why did I choose Bob Evans? The reason goes back to twenty years prior.

While we were students at Ohio University, Karen lived during her freshman and sophomore years in Jefferson Hall, which included the dining hall where I worked as "Mike the Milkman," running the cafeteria's milk machine.

Also living in Jefferson Hall at the time was Debbie Evans, the daughter of a man named Bob Evans, who was famous in Ohio for marketing sausage. We never could have predicted that Bob would continue expanding his business to include a line of food found in grocery stores and also approximately 500 restaurants found in eighteen states today. I remember talking to Mr. Evans a couple of times and even helping him carry Debbie's items up or down the staircase as I hung around the dorm's lounge waiting for Karen.

Little did I realize that the connection made in a dormitory at OU would lead to a personal challenge many years later. As I mentioned, each of the two recruiting squadrons I commanded included about forty offices and two MEPS. I tried to visit each office at least once every three months, so I did a lot of driving. A lot of driving led to a lot of boredom.

While driving to offices in my second squadron one day, I passed a Bob Evans restaurant and stopped to eat there, just as I had many times before. As I paid my bill at the check-out counter, I noticed a stack of brochures which included a map showing their locations. I took one to find other locations within my squadron's area, plus it was a pretty good map. I came up with an idea only a really bored pilot trapped in a cheap government car could conceive. *Why not see how many Bob Evans restaurants I can dine in during my travels?* So began my "mission" to stop and eat at *every* Bob Evans restaurant within my squadron's boundaries. Why? Who knows? It was just something to keep me occupied as I drove solo.

My mission was not as easy as you'd think. The 3552nd included four states, all of them in the backyard of the Bob Evans empire. But I continued to locate more restaurants as I drove around Ohio, Indiana, West Virginia, and Kentucky. My self-imposed rule for checking off a restaurant was not only stopping at the restaurant; I also had to eat there. As I mentioned my "mission" of visiting all of the restaurants to the occasional store manager, my fame evidently grew within the corporation, something I didn't realize until later. All I knew at the time was the food was good and reasonably priced, and it broke up the monotony when I pulled out of traffic for a while and ate a meal.

On those trips to visit the recruiting offices, my day didn't end when I checked into a motel for the night. In a squadron of approximately ninety officers, NCOs, and civilians, there were always effectiveness reports (ERs) to complete, both by proofreading what another person wrote and adding my endorsements and reports that I wrote because I was the person's rating official. This was difficult since all of these reports usually had to be sent to a higher endorsing officer, sometimes a one- or two-star general. Thus, they had to be as perfectly written as possible. I always relied on Karen, my editor, before I sent anything in.

Writing reports was hard enough to do in my quiet, well-lit office back at our squadron building. It was much more difficult to do in a motel after a long day's drive to multiple offices many miles apart. The Air Force didn't permit us to stay in the swankiest places available, so, normally, the motel I selected didn't have the nicest accommodations or equipment—like sufficient lighting and showers. Most of the places I stayed in cut costs by using low-wattage light bulbs and water-restricting shower heads. But I'd played this game before.

I learned from my year in New York to carry an "on-the-road survival kit" in my government car, a small cardboard box containing a couple of 100-watt light bulbs and a pipe wrench. The reason for the light bulbs should be obvious… but you might wonder about the pipe wrench.

I carried the box into the room and replaced any bulbs around the desk area with my 100-watt bulbs so I could see well enough to read and write the effectiveness reports I carried in my briefcase. Then, I'd check the shower to see if there was a trickle coming from the shower head or if it were a normal pressure head. If it was a flow-restricting shower head, I'd use my pipe wrench to remove the head completely. I stored the offending bulbs and shower head in my box, which I placed directly in front of the door of the room.

At checkout time the next morning. I replaced everything as it had been. Placing the box in front of the door ensured I would remember *my* bulbs and not forget to replace the shower head. In the three years I traveled across both squadrons, I never left a bulb behind or failed to replace a shower head.

After I'd been at the squadron for just a few months, one of my staff NCOs came up to me to share a story from my high school days. *How did he know that? He is far too young to have been in high school when I was, and he didn't even grow up in Ohio.* After this happened a few more times, it was driving me crazy. I had to solve the mystery.

For the next few days, I kept an eye on him and saw him appear to sneak out of the office one morning. I followed him at a distance, keeping out of his sight. He walked toward the building's cafeteria but made a right turn in another direction, proceeding into the Accounting & Finance section at the far end of the building from our offices. *What is this about?*

I gave him a little time to reach wherever he was going, and then I went through the same door he'd entered, stood at a distance, and listened to see if I could hear him talking. Sure enough, I heard his voice and the voices of two others coming from behind an office partition. I hung back until he finished talking and hid behind another partition until I saw him leave.

I then walked to the location where he'd been talking and recognized two of my high school classmates, Glynda and Bonnie, seated in the area. They immediately knew they'd been caught, and we all had a good laugh about what they'd been up to with my NCO. When I returned to my office, I let him know I figured out where he was getting his Intel on my high school indiscretions.

As our time on base stretched to more than a year, we still hadn't found a home to purchase. We attended every open house in Tipp City, but none of the homes seemed right for us. We wanted a quiet street, at least a two-car attached garage, and a decent-sized yard to play in...Face it, we were picky. Since we'd lived in something like fourteen houses over the years, we knew what we wanted, so we decided to wait for a new development to start.

In the meantime, we continued to enjoy living on base. Karen volunteered at the Officers' Wives Club's base thrift shop once a week, where she found two pairs of small-sized, double-bladed ice skates for the girls. When the water in the Turtle Pond near the brick quarters froze, Karen taught Lori and Katie to ice skate.

When the weather warmed, the four of us rode bikes to the Duck Pond, with both girls in seats on the backs. We fed the ducks with corn we'd purchased at Saunders Seed in Tipp City.

On nice nights, I'd take Lori for walks along the nearby base golf course to see if we could find any stray balls that might have gotten lost by players during the day. I always carried a couple of golf balls from previous walks in my pocket, and, if it looked like we wouldn't find any new ones that evening, I'd divert Lori's attention and roll one ahead of us for her to find.

As I neared the end of my second year as the commander of the recruiting squadron at WPAFB and my third year in recruiting, I knew I was ripe for another move since most command positions in the Air Force are two-to-three years in length. Since I'd done a good job at both Syracuse and at Wright-Patterson, I was asked by our group commander at Chanute AFB if I'd like to extend for a third year at Wright-Patt. That would mean four total years in recruiting. Decision time.

As was my habit, I looked for another possibility. I found there was still only one other Air Training Command position that I had a chance of getting that would keep us from moving. That job was the Professor of Aerospace Studies (PAS) at Wright State University. The university is located almost right outside the Wright-Patterson AFB gate, and its PAS would be replaced in the coming year. If I couldn't get that, I knew I'd either stay in recruiting one more year and then be assigned someplace else, or I'd choose to retire.

The decision was pretty black-and-white. If I accepted the offer to extend for another year in recruiting, I'd miss the opening at Wright State and would be open for a worldwide assignment the following year. The decision wasn't hard. Now, how could I make certain I would get the PAS position?

Without once more going into a list of possibly incriminating details (How long is the statute of limitations?), let's just say a *very good* friend of mine was the Vice Commander of Air Force Reserve Officer Training Corps (AFROTC) headquarters at Maxwell AFB, Alabama. And, without going into a lot of details, he owed me a VERY big favor.

I might add that I actually was highly qualified to assume the position, as I was one of only a few individuals who had instructed and evaluated the other two commissioning sources—Officer Training School (as a flight, squadron, and group commander), and the Air Force Academy when I inspected its flight training program. Also, assigning me as the PAS would save the Air Force money since I was already assigned to Wright-Patt, and no personnel move would be needed.

In short, I applied for the position of Professor of Aerospace Studies at Wright State University and was accepted. This was a major milestone for my family, as Karen and I decided this would likely be my last assignment before retiring. By this time, Karen and I knew that, as much as we loved the time we lived in Texas, we enjoyed being in Ohio and near relatives and old friends. So, we decided I'd retire at the end of this supposedly three-year assignment.

Now, with my future in the area of Tipp City locked in, Karen and I planned for a house to be built in a new subdivision in Tipp City. Lori completed kindergarten at a school in Fairborn, but once we moved to Tipp City, she and Katie could stay there through their public-school years. Moving to Tipp City would also allow Karen to look for a position teaching school and allow me, eventually, to look for a civilian post-retirement job.

I was excited to make the move to Wright State, and, although we enjoyed our years living on base, we all looked forward to building a new home in my hometown.

Our daughters, Katie (L) and Lori (R)

"Flying" with Katie in the AF Recruiting Service's mini-jet

Chapter 8

Back to School Once More

Going from my position as the commander of a recruiting squadron to the dual-titled *Commander of AF Reserve Officer Training Corps Detachment 643* and *Professor of Aerospace Studies* (PAS) at Wright State University in July of 1988 was an easy move of about four miles from the base to the university. But, of course, the new job meant attendance at *yet another* school in order for me to get checked out for the position. To this day, I find it interesting that I had to be "checked out" for every position I held in the Air Force except for the one position for which I knew absolutely nothing—the commander of a recruiting squadron. I have no explanation.

My check-out for the AFROTC position was a two-week course for new, incoming AFROTC commanders at the headquarters of Air University at Maxwell AFB, a place I knew well from my one-year assignment as an Air Command and Staff College student only five years earlier. The class was small, with only twelve lieutenant colonels and colonels. After the many years I spent as an instructor in different training positions, I thought the school was easy. Since the other future PASs were fun to be around, and there was little pressure as we were all senior field grade officers, I had a pleasant time sitting in class and completing assignments. Soon it was back to Ohio and into the new position.

I'm sure it is blasphemy for a military pilot to say that a non-flying assignment is among their favorite assignments, but, in my case, that is true. I really enjoyed the interaction of the students with the instructors and staff as both an OTS instructor and commander and as the professor of aerospace studies at Wright State University. I had a great deal of student interaction as an instructor pilot, but since there was so much pressure, especially on the student pilots, there was not a lot of time to relax with them.

Living on base at Wright-Patt made for an easy commute to and from work at Wright State, and my time as an OTS supervisor gave me immediate insight into the training procedures and course requirements of AFROTC. I had a professional staff of captains and noncommissioned officers, and I instantly felt at home and on top of the role I had walked into. Our cadets were students at Wright State University, the University of Dayton, or one of the other ten universities that our detachment included. They were all top-notch students, eager to learn and then enter the active-duty Air Force.

There was another bonus. Unlike recruiting service and flying assignments, AFROTC had regular hours and almost *no* travel away from home, so I spent more time with my wife and daughters, including time spent planning our new house. As soon as lots went up for sale in the new subdivision in August, we selected one and worked with a builder on the blueprints for our dream home. Construction began shortly after.

It quickly became obvious that our builder was better at *talking* about building houses than actually *building* them and staying on a schedule. Progress was so, so slow, but I foolishly thought the pace would pick up as the weather warmed in early spring. When it finally looked as if the end was near, I notified the base housing office of our impending move from base housing. Unfortunately, I had to delay the date for our departure so many times that the housing office finally declared, "That's it. You have to vacate. We have a new family scheduled to move in on April 1."

We informed the builder, and the house was completed, except for a few items that could be added after we were living in it. Therefore, after nearly a year of planning and construction, we moved out of base housing and into our new home with the help of my brother Steve and my nephews. As a matter of fact, we were the first family to move into this subdivision. At least the nearly daily trips from the base by Karen or me to check on construction came to an end since we were now occupying the house.

There were many changes in our lives that year. Lori transferred to the Tipp City schools after spring break of her first-grade year, and Katie was enrolled in a local preschool to finish her year. It had taken longer than we'd expected to complete the house, but we were getting settled, and the future looked bright.

At first, we missed all of the good things about living on base (one of them being having neighbors!), but we were happy to establish roots in a location where we expected to stay for many years beyond my Air Force retirement. Also, it was wonderful to have a large house and live in the same town as my parents and other family members and to be back in my hometown with many of my old friends.

While I would miss the easy four-mile commute to the detachment from the base, I suppose that after all the years of moving around, I just valued stability. I got used to the new twenty-five-mile commute. Karen, however, was hired to teach in the Tipp City schools in the fall, and she had a one-mile commute, so I guess that evened things out.

As the detachment commander, I could choose to teach any level class. I chose to teach the freshman *Introduction to the Air Force History and Customs*. Some PASs preferred to teach the senior class, but I thought teaching the freshmen gave me a better chance to meet the incoming cadets, learn their names, and keep me in the instructor role. I'd occasionally teach the sophomore through senior classes, too, just to keep up with how the cadets were progressing. I can say, however, that I *never* taught anything resembling a drill and ceremonies class.

The AFROTC detachment pretty much ran smoothly, without drama, and both the cadets and the staff were recognized as one of the top units in the United States. We received an excellent rating during our Inspector General inspection, and the performance of our NCOs and officers was held up as an example to other detachments within the region. Thanks to the cadets' efforts alone, our little detachment was elected to serve as the national headquarters for the Arnold Air Society (AAS) honorary organization—a truly BIG deal.

Due to the efforts of our national Arnold Air Society cadet staff, our AAS squadron was invited to attend the Air Force Association's (AFA) 1989 annual meeting in Washington, D.C. The AAS national staff from Wright State University was honored at the AFA conference for their leadership over the nation's 149 AAS squadrons, and I received the AFA Medal of Merit as their national advisor. My biggest task at the conference was herding thirty-plus cadets through the Washington Metro system while we were all attired in our formal Air Force mess dress uniforms. I'm happy to report that we all made it to the conference and back to our hotel without losing anyone. For that alone, I deserved the medal.

One way we undoubtedly differed from other detachments across the country was that we often held staff meetings at the local Bob Evans restaurant across the street from the Wright State campus. In time, some of our cadets discovered our hangout and started to join us there when they had time. We ran a truly professional detachment, but I thought it was good to show the students that we—those of us in "the real Air Force"—were just nice people with a difficult job to do in the real world.

Throughout my military career, I was known for some unorthodox behaviors I either exhibited or introduced to my unit. When I discovered that none of my staff had heard of a game we played in my previous squadrons called "Three Coin Poozeman," I had to change that. The game originated in early flying units as a way to have someone pay for drinks at the bar, but over the years, it just became a game to, hopefully, get someone else to pay for anything.

The concept is simple. Each player holds three coins behind their back and brings out any number or none of the coins in their right hand. The person who correctly guesses the total number of coins in the circle of right hands wins elimination from the game and is out of danger of buying. The last person left after all others are eliminated is the loser, who then buys for the other players. There are many variations and weird rules that have evolved over the years, but those are the basics.

Our staff would play Three Coin Poozeman for soft drinks once a day. The loser had to not only pay for the soft drinks but go to the machine on the first floor and deliver them to the other players. Our variation was the cans could be delivered in any manner the loser chose—so long as the structural integrity of the can was intact. This we called "the structural integrity rule." If the rule was violated—i.e., the can exploded or sprung a leak—the violator owed the group a second round of soft drinks at another time.

Structural integrity was tested and often violated in any number of ways, as each loser tried to be more creative in their delivery method. It wasn't uncommon to hear cans being rolled down the hallway into offices or to see cans fly into the air outside our window and land in a puddle, only to be delivered in a less-than-pristine condition and wet or muddy. As long as there were no leaks or exploding cans, all was fair. As I said, we were a truly professional staff…most of the time.

Someone at AFROTC headquarters must have liked me—or liked our detachment's performance—because I was selected to be the guide for an unusual jaunt. I later found out it was the big guy himself, the commander of AFROTC, General Ellis. I got a call from my immediate boss, a colonel at the AFROTC regional headquarters. "I received a by-name request for you to escort a group of ten Royal Air Force (RAF) student pilots across the United States as part of an exchange tour with Great Britain," he began.

What? What's this about? I wondered as the colonel continued.

"The tour will be overseen by the British Embassy, the American State Department, and Department of Defense officials." I was to accompany the group, along with their liaison officer (a squadron leader with the Royal Air Force rank equivalent of an Air Force major), make certain things went smoothly, and iron out any potential complications before they became actual problems.

I flew to Washington, D.C., commercially on July 24, 1989, where I was met at the airport by a Department of Defense (DOD) civilian and a large 18-passenger van. The Brits arrived about two hours later, and the DOD representative took us to a nice hotel and said, "I'll pick you up tomorrow morning for a tour of Washington sites."

That night was an opportunity for the British students, their squadron leader, and me to get to know one another. The student pilots were about 22 years old—the same age as my former UPT student pilots—and let's say they were *extremely* enthusiastic and excited, for lack of less diplomatic terms. In other words, they were fun, and right away, I liked being with them.

The first night at the hotel went smoothly, and the next morning the van picked us up for a day of sightseeing. To describe this as a VIP tour would be an understatement. Even I wasn't prepared for the way we were treated.

Everywhere we went, we were met by a waiting tour guide and given VIP access to the sites. For example, when we pulled up at the Washington Monument, a National Park ranger ushered us around the waiting line of tourists to the elevators and gave us what amounted to a private tour. Another highlight was a tour of the Pentagon, which I myself had never been inside. We went to the White House, too, but we weren't invited in by President George H. W. Bush or First Lady Barbara for tea and crumpets. Darn.

After two days in the nation's capital, my merry band of Brits and I were off to Colorado Springs, Colorado, for tours of the Air Force Academy and the North American Air Defense Command (NORAD) Cheyenne Mountain Complex, which both they and I found to be amazing. Of course, I had heard about the Cheyenne Mountain Complex while growing up during the Cold War, but it was even more amazing than the stories I had seen and heard. It is a totally nuclear-hardened and self-sufficient command and control center underneath a granite mountain, where the entire command center is built on massive shock absorbers. My preconceived vision of what it looked like didn't come close to its reality. This was a side of the Air Force unlike anything I'd ever seen, even though I'd had a tour of the SAC command and control center at Offutt AFB in Nebraska while I was flying B-52s.

Next stop: four days at Williams AFB in Phoenix, Arizona. This stop was especially interesting to the Brits, as it provided a chance to compare British and American pilot training techniques and procedures. All of the British pilots got a flight in either a T-38 or T-37 aircraft, and they had plenty of time to talk flying with the American students. The British squadron leader and I both decided it was far too hot (at 100+ degrees) to don flight suits and go sweat in an airplane for an hour and a half, plus time on the ramp for pre-flights and engine start-up. No, thanks.

Since we were heading into a weekend, the base supplied several student pilots and a public affairs officer to show the Brits the local area and its night life while I remained around our BOQ and relaxed. I'd inspected the UPT program at Williams a number of times while I was in ATC Stan/Eval and didn't need any night life other than a TV, a pizza, and air conditioning.

Our next stop was Holloman AFB near Alamogordo, New Mexico, for a three-day stay to tour a fighter base. The Brits were offered a chance to fly in a fighter aircraft, and, once again, the squadron leader and I elected to pass on flying and instead went to the Officers' Club for a nice, quiet breakfast and coffee—and air conditioning, of course.

The last stop before our visitors returned to Washington, D.C., for their flight home was conveniently—and not by accident—Wright-Patterson AFB. There was a little more to the timing of this stop than they knew, as my high school graduation class was having its 25th-anniversary reunion on the same weekend we'd be at Wright-Patt. When I saw the original itinerary, I realized we were scheduled to be in Omaha, Nebraska, on the day of my reunion, which I was helping to plan. So, I contacted my DOD liaison at the Pentagon a month before the trip, and they kindly reversed our itinerary, so Wright-Patt was placed at the end instead of at the start of the tour.

While I was at the reunion, my boss, the colonel, was happy to meet the Brits at the Officers' Club and regale them with how the Air Force was "back in the day." The next day, I picked up the Brits at the base in a government van and drove them to my house for a party, complete with a number of girls who were friends of the daughter of one of my high school classmates. Everyone had a good time, and we all ended up at the classmate's house for a pool party, as earlier, I'd told the Brits to bring swim trunks to my house.

Everybody was having a great time...until we weren't. Luckily, a neighbor from my high school days was on duty as a deputy sheriff that evening, and that's all I'll say about that. Let me just say that all turned out well, and the Brits left for Washington, D.C., and the trip back to England the next day. Actually, all of them left for the trip home except the one student who was in the Wright-Patt hospital recovering from a broken leg. However, he recovered just fine, an international incident was avoided, and the DOD and the British Embassy both still loved me. End of that story.

I couldn't have been happier serving in the AFROTC detachment as I contemplated what I'd do when I retired and became a civilian after 23 years in the military. But, and there's always a but, AFROTC headquarters decided that after two very enjoyable years and an outstanding inspection by their headquarters, I could do the Air Force more good assisting the new, incoming colonel and his staff at our regional headquarters on base.

I suppose I should have looked at the move as a step up, but I had mentally plotted out my final year in the Air Force, and this wasn't in that plan. I'd much rather have spent my last year on active duty at my detachment with my good friends and the students. However, as the saying goes, "The needs of the Air Force come first." Darn.

It was fortunate that I was staying at Wright-Patt instead of transferring to another base, of course, but going from being the commander of one AFROTC unit to being the vice commandant of the entire ten-state Ohio Valley Region was not that great. The region encompassed ten states, 27 university detachments, and Air Force Junior ROTC detachments in 43 high schools. That sounds like a step up in responsibility, and it was, but I wanted to finish my final year in the Air Force in a job I enjoyed and where I was in charge. This wasn't it.

I considered the situation. I survived a year in Vietnam while being shot at every day, so I knew I would survive a desk job I didn't particularly care to have. I knew that AFROTC headquarters meant the vice commandant assignment to be a recognition for my unit's outstanding performance, but I also knew I'd miss the cadets, the Detachment 643 staff, and the relative freedom of running my own program. And I did. I missed that a lot.

It was clearly time for me to retire from my beloved Air Force.

I know I've said this before, and I repeat it often to this day: The families of military service members are the true heroes and heroines of the military. They serve and sacrifice just as the service member does, sometimes more, but without recognition or compensation. Karen had been the perfect Air Force wife through many hardships. She moved with me fourteen times, not to mention the year I was gone in Vietnam and the nine temporary duty assignments—each lasting one to three months, which I was sent on for various reasons. She deserved to finally pursue her own career path.

The children of service members also serve. My daughters and Karen had to attend events where both civilian parents of classmates were in attendance while I was off at some Air Force school, assignment, or temporary duty. By the time Lori was five years old, she had moved five times. Katie had moved three times by the time she was three. It's impossible to maintain long-term friendships under those circumstances.

So, in went my papers to retire. I loved my time in the Air Force, and I knew I'd miss it, but it was time for me to defer to the wants and needs of my family. And, truthfully, I didn't want to move again.

My retirement ceremony in August of 1991 was the very best I've ever seen. In five years, I had served in three separate units since arriving at Wright-Patterson AFB in 1986, and none of them had forgotten me. That alone was worth celebrating, but they all combined to give me one heck of a retirement celebration.

One of my pilot training classmates, Bill Orellana, had been promoted early to full colonel and was the base commander at Wright-Patt, and he agreed to officiate at the ceremony. The event was not held in a small conference room on base, as is often the case. Nope. It was held at Wright State University's Nutter Convocation Center in a large special-event room. I was honored that my friends in my former unit, AFROTC Detachment 643, not only hosted the event but provided the honor guard and helped plan the event.

My recruiting squadron at Wright-Patt also helped with the ceremony, and many of my recruiters, staff, and officers participated in or attended the ceremony. It was a wonderful although bittersweet event, and presentations were made not only to me but to Karen and to my daughters. It seemed like everyone I knew was somehow involved in the program, from my parents, brothers, and sister, to the mayor of Tipp City, who presented a proclamation declaring my retirement "Lt. Col. Mike Jackson Day in Tipp City." I was blown away when a corporate representative from Bob Evans restaurants stepped up to present me with a *Bob Evans Gentleman Farmer* certificate gifting me one square foot of Bob Evans Farms. Everyone involved made it a beautiful, touching ceremony. It couldn't have been a better culmination of my Air Force career.

So, now I was Mike Jackson, "retired Air Force guy," but I didn't stay retired for long. As a large amount of my military career involved education and instruction, from being an instructor for the Vietnamese Air Force in Vietnam to instructing student pilots, officer trainees, and AFROTC cadets, teaching seemed to be a logical route. And, since Karen was now teaching at Tippecanoe Central Intermediate School and Lori and Katie were still in grade school, our schedules would match perfectly for school vacations and summer breaks. Theoretically, it seemed like a *great* idea. Theoretically.

One week after I retired, I began teaching history classes at Milton Union High School in a small town near Tipp City. West Milton was Tipp City's archrival in sports when I was in high school, but Tipp had grown much larger than Milton due to an interstate highway that, for better or worse, now passed through

Tipp, resulting in significant growth to my little town "where everyone knew your name."

What I hadn't taken into account when I applied for a teaching position was that I would no longer be in charge of anything. Furthermore, my neck and back pain issues wouldn't hold up well to standing all day, and the three-minute breaks between classes didn't allow time for me to stop and rest when needed.

Furthermore, I could no longer stop somewhere on the way for breakfast—a habit I'd enjoyed since moving back to Ohio—since I had to be at school earlier than the local West Milton restaurant opened, but that was really minor compared to my health issues. On the positive side, I enjoyed getting to know the students. I could understand why Karen chose the career she did.

After just a few weeks, I realized that my public-school teaching career wasn't going to be long-term. When I'd ended up in traction on a part-time basis for three weeks of my last few months before retiring, I should have taken that as a warning sign. The principal couldn't have been more understanding and sympathetic to my situation and allowed me some leeway with appointments at the Veterans Administration hospital as needed.

Once again, the Jacksonian luck came through. I ran into Jim Gallagher, a friend from my days at Wright State University. When he asked me how teaching was going, I told him I enjoyed the students, but I was having pain issues that would undoubtedly shorten my teaching career.

Jim was the ultimate networker. He told me that the National Aviation Hall of Fame (NAHF) organization in Dayton was searching for a new executive director with a background in aviation history and management. The next thing I knew, I was invited to lunch at the Dayton Racquet Club in downtown Dayton to meet the NAHF Board chairman, Neil Pugh, and president, Don Vanderkarr. While talking to Neil and Don, I realized that the executive director position would be a great match for my background and skills—but it would also be an extremely challenging job.

The NAHF was formed in Dayton about thirty years earlier, in 1962, and it annually inducted four aviation pioneers with a prestigious enshrinement ceremony. However, up to that point, the only physical presence of the Hall of Fame for the remainder of the year was a row of portraits of the inductees hanging on a wall in the Dayton Convention Center. What the Board of Trustees of the NAHF envisioned was building their own facility and initiating full-time programs to better recognize the pioneers of American aviation.

I should mention that at that point, there was no NAHF staff. It had previously consisted of only one individual operating from a small, donated office in the Dayton Convention Center, but that person had left. And, the organization was in debt. The NAHF Board offered me the job. Hey! Who could pass up a challenge like that? But first, I had to resolve my teaching situation.

I returned to West Milton and told the high school principal and superintendent what I wanted to do. They understood my situation, and we agreed I'd finish the semester before leaving for the NAHF. After one of the shortest teaching careers in history, I was off to become the executive director of the National Aviation Hall of Fame. In hindsight, I probably should have asked what my salary would be.

WSU AFROTC staff playing "3 Coin Pooseman"

British student pilots during their tour of U.S. bases

Family photo – Katie (left front), Lori (right front), Karen, and Mike

Katie and Lori in Mike's former supersonic classroom, the T-38

Retirement ceremony with Col. Bill Orellana

Lori, Katie, & Mike received autographed photos from the Thunderbirds

Chapter 9

Tiny Office to Robust National Organization

I'd describe my first day at the National Aviation Hall of Fame (NAHF), on February 10, 1992, as "interesting," if nothing else. The office, and I use the term loosely, was a small, storage-closet-sized room located within the Dayton Convention and Visitors Bureau's suite of offices in the Convention Center in downtown Dayton, Ohio. It was just big enough for two metal desks and two bookcases. There was one desk phone and an older Apple computer which I had no idea how to operate.

And, just to make things more challenging, the motor of the center's elevators was located in a large closet in the back of the office. Every time the elevators went up or down, the sound put a serious crimp in any in-person discussions I held or phone calls I might place. But the office was free, as it was donated to the NAHF by the Dayton Chamber of Commerce. Free is good—very good—when you have no money, so I appreciated the gesture. It provided a place for me to begin building the organization.

On that first day, the kind staff of the Convention and Visitors Bureau, who remain friends to this day, showed me around the building. Over the next couple of days, some highly-involved, dedicated members of the NAHF Board of Trustees showed up to help me learn the ropes. A number of them were founding members who remained involved with the organization since its founding as the Aviation Hall of Fame in 1962. They provided the institutional history, as well as their full support, and I was lucky to have them.

I knew after just a few weeks that this job wouldn't be anything like the ones I'd had before when the phone rang. "Hi, Mike. This is Alan Shepard. I'd like to see if you have any enshrinement photos from the 1977 ceremony." I didn't immediately realize who the Alan I was talking with was and faked my way through the conversation until I realized I was speaking to Alan Shepard, the first American in space, who later walked on the moon. Yes, this job was going to be *very* interesting!

My initial tasks were two-fold: increase the NAHF's financial security and expand its activities that summer to more than just a one-night induction ceremony for the aviation pioneers who were to be enshrined that year. Those two items gave me plenty to do, but I was excited at the possibilities. There were numerous problems to solve, but I've always liked a challenge. This challenge was huge.

The National Aviation Hall of Fame was definitely in need of an executive director and reorganization. I followed three executive directors in the six prior years, and the last director had been dismissed. Not much was done administratively in the recent past except by the involved members of the board. The organization had a significant debt, primarily a loan from one of our patrons who was also a valued member of our board.

During my first two months at the NAHF, I was the sole employee, and, despite occasional help from board members, I needed to find someone who could run the office while I tried to raise operating funds. I began a search for an administrative assistant and was lucky to find a recently-retired executive secretary who wasn't ready to fully retire yet. I hired Cecelia (Cee) John in late March, and we became an outstanding team.

Cee had worked for a major oil company. She was extremely organized and talented, and I was fortunate to hire her. She was also older than the other applicants, which I viewed as a plus. Her work ethic was unmatched, and she planned to stay for the long haul.

At that time, the NAHF board was overseen by the Dayton Area Chamber of Commerce, although the chamber didn't take an active part in its day-to-day operation. When the chamber's vice president heard that I hired someone who was 59 years old, he called me upstairs to his office, where he proceeded to tell me in a not-too-subtle manner, "You've made a big mistake, and I won't allow it."

I reminded him, also in a not-too-subtle manner, "I'm not some kid you can yell at. Less than a year ago, I oversaw a ten-state Air Force unit with a multi-million-dollar budget and hundreds of officers and enlisted personnel, a much larger organization than you're running as an assistant to the president of the local chamber. End of conversation."

I stood up briskly, and, as I approached the door, I added, "I won't put up with your language or attitude. Do not *ever* call me to your office again until you can treat me as an adult." Then I left.

I returned to my office and, without telling Cee exactly what happened, said, "I might not have a job anymore." I honestly didn't care. I had a retirement check, marketable skills—and a working wife.

The next morning, I got another call from upstairs, this time from the chamber president, asking me to come to *his* office. *Uh-oh*, I thought, *I've never been fired before.* As it turned out, my "discussion" with the vice president was never mentioned, and all the president said—in a polite way—was, "You'll be reporting directly to me from now on." Whew!

Fortunately for me, the chamber had a shake-up a short time later that resulted in major changes in its leadership and the separation of many of its organizations into independent units, including the Visitors' Bureau, the Air Show, and the NAHF. From then on, I had a great relationship with the new vice president and, eventually, their new president. All of the individual units prospered, too.

Prosper is what the NAHF began to do. With Cee running the administrative side of the organization, I focused on raising money to repay our debt, pay salaries, and expand the organization from a once-a-year ceremony into a national presence—with an eye toward building a new permanent headquarters, one away from the convention center.

My fundraising efforts began slowly, but as I gained experience, things picked up speed. Soon, I was able to pay off all debts and actually be allowed a small expense account. Until then, I paid for business meals and gasoline for official trips out of my own pocket—from a smaller salary than I'd had as a first-year teacher.

At the same time, the increase in revenue allowed for our annual enshrinement ceremony to become a more professional, more entertaining event, which led to a bigger audience. As word spread and our publicity increased, people started to take notice.

Now that my family was settled in Tipp City and I began to renew friendships with people in town, I wanted to become involved with local organizations. After investigating possibilities, I decided to volunteer with the Tipp City Fire Department, a completely volunteer unit. I found the other volunteers to be a great group of public servants, and I thoroughly enjoyed the training and the actual firefighting. Washing fire trucks after a run wasn't the greatest fun, but I recognized it as a necessary task, and the department took pride in their equipment and their performance.

Another thing my stable home life allowed me to do was examine the contents of boxes we'd been moving from one house to the next. Although I was wounded during a Viet Cong rocket attack in May of 1972, I purposely didn't follow up with the paperwork for my medical issues until I retired from the Air Force, as I feared I'd be taken off flying status or, worse, medically discharged or retired. Once I retired, I felt free to gather my records and witness statements to apply to the Air Force and to the Veterans Administration (VA) for any disability compensation and awards I might be due.

After updating my medical records and obtaining written statements from my squadron mates from Vietnam, I began the slow process of applying for a Purple Heart, with encouragement from a kind doctor at the VA. The statements supporting my injuries were easy to obtain since my injuries during the rocket attack were pretty memorable, and some of my fellow O-2A pilots flew my scheduled missions until I recovered enough to return to flying. A Purple Heart medal was approved on November 16, 1992, more than 20 years after the injuries, but I couldn't complain. It was my fault that I'd kept the medical issues as hidden as I could while on active duty in order to remain on flying status and complete my combat tour of duty.

The Air Force stated that the Purple Heart medal should be presented in an official ceremony and suggested contacting Wright-Patterson AFB to arrange for a medal presentation ceremony to be held at the base. At this point in my life, I didn't want to make a big show of the presentation, so I asked if the medal and certificate could be presented at the local American Legion post during one of their meetings. I was told that was acceptable.

A high school friend of mine, Gene Chaney, was the post commander, and he was happy to present the Purple Heart to me during a Legion meeting in January of 1993. Gene joked, "This is great! A former Army enlisted man is presenting a medal to a retired Air Force officer." It was a low-key ceremony, and I was happy to receive the medal, although I knew I'd probably never wear it on a uniform since I was retired.

I thought that would be the end of the medal story until the city manager asked me to stop by his office a few weeks later. Since he was a friend of mine and we talked on a regular basis, I didn't think much about his request. As it turned out, he had read an article about the presentation in our local paper, the *Tipp Herald*. After some small talk in his office, he congratulated me on the medal, and I thanked him. He then asked, "Are you disabled from the injury you received?" I instantly knew what was coming.

"According to the VA, I'm eighty percent disabled."

"And you're on the fire department?"

"Not anymore?" I asked, knowing full well what he was going to say next.

“Not anymore.”

I knew I’d been caught, and he did, too. We both got a little laugh out of the exchange. In an instant, I became Scoop Jackson, the *former* Tipp City firefighter. Oh, well, it was fun while it lasted. Maybe then the assistant fire chief might understand why I kept taking my helmet off at fire scenes and during training because my neck hurt from its weight.

Now that I’d managed to stabilize the NAHF’s finances and survived my first enshrinement ceremony, it was time to focus on the board’s third goal: find a way to fund and build a separate building that would serve as both our administrative headquarters and as a learning and research center focusing on the lives and contributions of our enshrinees. Easier said than done. It would take four years from my first day at the NAHF until a decision was made on location.

In March of 1994, several Dayton area museums, including the NAHF and other members of the Ohio Museum Association (OMA), were to host the OMA’s annual statewide conference. Many meetings were conducted to plan the various events, venues, and sessions for the conference. Most of these meetings were held at the Dayton Museum of Natural History (which later became the Boonshoft Museum of Discovery). Judy Chovan, the museum director, and I had become pretty good friends by this point since I had spent a lot of time at her museum as I studied their exhibits, so I was familiar with the museum’s layout, including the office areas.

The conference started early one morning. After I’d spent hours in seminars and touring the exhibit tables of various OMA members and organizations supporting the museum trades, I’d pretty much reached the limit of my ability to attend seminars and just stand around and talk about exhibits and educational programs. In other words, I was really bored. Not a good thing for the staid and professional OMA, as it turned out. Hey, I was still an Air Force pilot in my DNA, and I was not above having a little fun…at someone else’s expense, of course.

Since I was relatively new to the museum crowd, other than to the folks in the Dayton area, I decided to adopt a new identity and have some fun with the people manning the exhibitor tables.

As I mentioned earlier, I was familiar with Judy's office areas, so I knew where our Dayton host group kept blank conference name tags and also the location of their computers and printer. While everyone else continued to attend seminars (which I should have been attending), I quickly made a new official name tag for myself and replaced Mike Jackson, Executive Director, National Aviation Hall of Fame, with Jack Taylor, President, Bunch of Bones Museum. Since all of the male attendees were dressed in coats and ties, I looked professional, indeed, as I headed back to the conference.

Under my new identity, I circulated throughout the museum exhibits and vendors' tables. I hadn't been very active in the museum community up to this point, so no one outside of the planning committee knew who I was, except for a few attendees who were sequestered in the sessions I was supposed to be attending. I did swear one exhibitor to silence. That was Chris Smith Perkins, a high school classmate and good friend who possessed a similar sense of humor.

The first table I visited was a display for an exhibit construction company. I signed their visitors' log as Jack Taylor, Bunch of Bones Museum. Until the company representative asked me, "What does the Bunch of Bones Museum do?" I really hadn't thought much about it.

Out of the blue, I responded, "The Bunch of Bones Museum will be a new, interactive, hands-on facility that teaches children and adults about the skeletal structure of all mammals." *Not bad*, I thought, for a spur-of-the-moment museum concept—and one that included many of the catchphrases popular with museums of the 1990s.

Jack Taylor moved on to the next table, leaving the previous exhibitor scratching her head about the mission of the Bunch of Bones Museum. Little did she realize that I was just as clueless as she was.

By the time I'd signed in at the second exhibitor's table, the mission of the Bunch of Bones Museum was starting to come into focus. At least as much focus as an imaginary museum with an impostor as the director could muster in about sixty seconds.

"So, what does the Bunch of Bones Museum do?" asked the unsuspecting exhibitor at the second table.

"Well," I said, "we hope to fill a room with as many bones from as many animals as we can find and then have children, school groups, and adults enter the room and either try to rebuild a specific animal skeleton by identifying the bones of that animal or use bones from different types of animals to build their own unique animal."

By the time I finished my impromptu description of the Bunch of Bones Museum's mission, I couldn't identify what I saw—disbelief, fear, or shock—in the eyes of the exhibitor.

I was on a roll, so I moved to table number three. Once again, I described the mission of the Bunch of Bones Museum to the exhibitor but with much more detail this time. The same reaction as at table two, but I noticed some of the exhibitors had edged to the rear of their exhibit area, farther away from me. I mean, farther away from Jack Taylor.

Before long, I had found the perfect missing part of my pitch. After explaining my fictitious museum's mission to table five or six, I added the best part. "We've arranged with local police and sheriff departments to call us if there's any road kill so we can add to our collection."

The *piece de resistance* to my sales pitch was my finale. "We're always on the lookout for human skeletons, too,'" and then my asking, "Would you be interested in hearing about our deferred giving program?"

As I scanned the large exhibit area, deciding where to head next, I noticed groups of people starting to gather and whisper, all the while looking at me. Only one person in the entire vendors' area had any idea who I really was—Chris Perkins—and she had promised not to expose me. I was loving it. But, before I could cover the entire room and before museum directors and staff members returned to the exhibit area from the sessions for which I was truant, my phone rang.

The night before the OMA conference, I attended a concert by my favorite band, the Moody Blues. I knew they would be staying overnight at the Crown Plaza, located directly across the street from my office in the Convention Center. I had asked a friend on the hotel sales staff, Teri Porter, to *please* let me know if any of the band members were moving around the hotel so I could rush there to try to get a photo with them. I had my Canon Sure Shot in my coat pocket, just in case. As fate would have it, just as Jack Taylor and the Bunch of Bones Museum were hitting their stride at the conference, Teri called to tell me that some of the band members were in the top-floor dining room and would be leaving the hotel in their buses after that. Time to leave the conference!

I drove the short distance from the museum to the hotel and waited in the dining room lobby until the Moody Blues band members Justin Hayward, John Lodge, and Ray Thomas left to meet the rest of the band in the lobby. I rushed into the elevator just as the door was closing and rode down with them, trying not to act like the crazed stalker fan I was. I guess the coat and tie I wore offset my stalker vibe, and they actually posed for two photos with me in the lobby. It was a good day for Mike Jackson…and for Jack Taylor.

Nine years later, in 2003, I was selected as a Distinguished Museum Professional by the Ohio Museums Association and given a nice trophy during a conference in Cincinnati. They introduced me as Mike Jackson, Executive Director of the National Aviation Hall of Fame, and as I walked to the stage to accept my award, I could hear people in the audience ask each other, "Isn't that Jack Taylor from the Bunch of Bones Museum?" I had to smile. The legend of Jack Taylor lives!

While the NAHF board members continued to explore possible locations for our building, I made headway on the funding, thanks to generous supporters in the private sector, corporations, and state and federal government grants. However, selecting the building's location proved to be divisive, both internally and geographically. Some sectors wanted it built within the city limits of Dayton; some wanted it at Wright-Patterson AFB at the Air Force Museum (AFM) complex; some wanted it in other towns or states. And, of course, political figures had to voice their opinions. It was not easy to reach a consensus.

After examining various possibilities within Dayton and beyond, the board decided to attempt to build the learning center at the Air Force Museum. That led to intense discussions between the board, the AFM, the State of Ohio, and the Department of the Air Force on whether it should be attached to the museum, near the museum, or on land around the museum. My feeling was that it should be a totally separate building on the museum grounds, away from the museum building itself. I had many reasons for that opinion, but as the executive director, I took directions from the board and had no vote. So, it was decided the building should be attached to the AFM between its first and second hangars.

After attending meetings with state and local entities, I traveled to the Pentagon with our architect, Steve Carter, and the WPAFB Judge Advocate General (an Air Force lawyer) to meet with officials from the Department of the Air Force. At one point, I was in a meeting with a four-star general, ten Pentagon architects and lawyers, our architect, and the Wright-Patt JAG, Colonel Jeff Owens. Yikes! But it all worked out—for the most part—since a compromise never makes *everyone* completely happy, particularly a group this large and diverse.

From July 9th to 12th, 1996, a final four-day meeting was held on base with representatives of the Secretary of the Air Force, the State of Ohio, and WPAFB. Those present included base and private civil engineers, Colonel Jeff Owens, Steve Carter, NAHF lawyers (including board member Bob Spayd), and me to work out lease arrangements at the AFM.

On July 19, 1996, a ceremonial groundbreaking for the NAHF learning center was held between hangars 1 and 2, although actual construction couldn't be scheduled until architectural plans were approved up the chain of command. I suspected all of the entities involved wanted a ceremonial, well-publicized groundbreaking as soon as possible so no one could change their mind.

During this time, my fundraising continued, allowing the mission of the NAHF "to honor America's air and space pioneers" to advance. Additional staff was hired, so we needed additional room to accommodate them. The Air Force Museum was kind enough to loan us office space in their presidential aircraft hangar, which was then located in a separate building but close by.

I was now tasked with planning and designing interactive exhibits, which I had never done before. The exhibits would focus on both the lives of the enshrinees and their contributions to the history of aviation. I was, of course, able to contribute to both the educational and aeronautical concepts in the designs. The process was all the more interesting since, by now, I'd gotten to know many of the enshrines personally.

With the help and advice of local area exhibit experts like Judy Chovan and my visits around the country to museums whose displays included interactive exhibits, a plan for the learning center began to take form. And, as plans for the exhibits took shape, I could approach potential donors for funding since I now had drawings to show. I began to think I actually knew what I was doing—or maybe I was just getting better at bluffing.

Things were moving along smoothly with the NAHF building plans and for me as the executive director. However, I couldn't put my service in Vietnam out of my mind. Like most Vietnam veterans, and veterans of the Vietnam War era in general, I am still deeply affected by the way we were welcomed—or, I should say, how we were *unwelcomed* upon our return home. We were the good guys, the guys and gals who answered our country's call, either voluntarily or through the draft, and entered the military during the Vietnam era.

Throughout our nation's history, there have always been individuals who favored and others who opposed the military and/or the conflict in which the nation was involved. That was true during the American Revolution and during the Civil War, as well as during other conflicts.

However, the war in Vietnam was different. It was brought into our homes every evening on the network news. We couldn't put it out of our minds. Due to biased reporting, the war seeming to drag on and on, and especially due to the actions of unscrupulous politicians, the public turned against the war. Politicians were the only people who voted to send troops to fight and die in a foreign war, but the public turned against the military members, the individuals who didn't get a vote on the conflict or how it was fought. The military became scapegoats for the war and the target for the wrath some of the public felt. And the media and Hollywood piled onto this unjust treatment, usually portraying a veteran of that period as mentally unstable, addicted to drugs, and, generally, someone to avoid.

Downtown Dayton, Ohio, is no different than any large metropolitan area in that it has its share of unemployed and homeless on the streets. Until I began working at the Convention Center, I had not been around many homeless. As I walked around downtown, I noticed many of the street people either identified themselves as veterans on signs they held, or they wore old military articles of clothing. I never gave any of these people money, but I did occasionally give someone carry-out food.

Over time, I began to recognize some of these people, so I'd say *hello*, and sometimes we'd have a brief conversation. It was during one of these encounters that I learned one of the guys I'd come to recognize was not only a Vietnam veteran, but he served at Camp Eagle at the same time I was there. That led to a "moment of truth" for me.

I discovered he had been turned down for so many jobs after his return from Vietnam that he had basically given up. I told him he seemed to me like the kind of person any company would be lucky to have, and he should apply to a couple of places I could recommend.

Another "moment of truth." He asked how he could apply for work since he had no phone number or address to put on an application. And, considering the way he was dressed, he felt he couldn't apply anywhere in person, assuming he could figure out how to get to the business location without transportation.

I eventually took this gentleman to my office after hours and helped him fill out an employment application using our office's phone number and mailing address, and I gave him some of my clothes for his interview. He got a job and was eventually able to get off the street. Wow!

This began a path for me that I was to follow for the next thirty-plus years: helping veterans, particularly Vietnam veterans, by helping them realize their contributions during their military service were worthwhile. And, to my surprise, the more I helped others feel better about their service, the better I felt about my own. It felt better to help fellow Vietnam veterans than just be angry about the public's actions upon our return.

This led to my being asked by employees of the Dayton Chamber of Commerce to take part in their pre-retirement presentations at the base to service members who were in the process of separating from the military. The Chamber of Commerce folks instructed the service members on how to apply for jobs and how to "civilianize" a resume. I offered my perspective as a military retiree.

The presentation was well-received, and the Chamber included me the next time, as well. Following that, the base Family Support Center asked me to continue. I shared my experiences and advised them of the resources available through various veterans' organizations, including the Montgomery County Veterans Services office, whose former director, Denver Combs, became and continues to be a good friend. Incidentally, Denver and I learned we were stationed at Hue Phu Bai in Vietnam at the same time. Small world.

In 1997, for my work with veterans, I was inducted into the Ohio Veterans Hall of Fame. I didn't even know such an organization existed until then, and any recognition was never on my mind when I began assisting veterans.

By mid-1998, the funding for the NAHF building was in place, the exhibits were being designed and built by the PPI exhibit company in Portland, Oregon, and it was time to leave my home of six years in the Dayton Convention Center and move all of the staff into our nearly-completed learning center at the Air Force Museum. We wouldn't have the interactive exhibits completed and installed for months, but we'd had temporary exhibits designed and built, with a plan to use them as traveling exhibits in the future.

On August 2, 1999, the new National Aviation Hall of Fame Learning and Research Center within the National Museum of the United States Air Force opened to the public with the temporary exhibits. Thirty-seven years after the NAHF was established as the Aviation Hall of Fame, the only congressionally-chartered aviation hall of fame, it existed no longer as just an office and a mailing address. The organization had a permanent, physical home.

By this time, we had added three additional members to our staff, and we'd become an organization to which other museums and learning centers looked as an example of how to grow a successful organization. The staff of various museums and non-profit organizations were actually coming to *me* for advice. Quite a change for the man who had no clue what the executive director's job entailed when he first entered that small office in 1992.

In preparation for our opening to the public, we recruited and trained a cadre of aviation-oriented volunteers to work on the exhibit floor. These individuals were committed to the NAHF's mission, and a few of these dedicated volunteers remained with the NAHF for more than twenty years. We couldn't have remained open as a learning and research center without our volunteers, and I remain grateful to them. One of our volunteers, Amy Marino Spowart, would remain for many years and, eventually, became the president and CEO of the organization.

About one year after we opened, one of our employees mentioned there was a volunteer who not only knew a good deal about aviation and our enshrinees but was also a talented writer. She was happy to be a museum volunteer, but she wanted to focus more on the people of aviation than on the machines. I made it a point to meet her during her next volunteer shift, and I was impressed with her strong interest in the NAHF's mission and her knowledge of aviation in general. She shared that she was working on her private pilot's license. That was my initial meeting with Tara Dixon Engel and the start of a long business relationship and a friendship that extends to this day.

Tara was eager to volunteer in our offices, so I put her to work in our research center as a volunteer archivist. I'd been told the truth about Tara; her writing skills were exceptional. I ended up hiring her as the director of the research center and, as such, asked her to research and write development grant requests and strategic plans for me to follow on fundraising calls. Between her skills, Cee's administrative expertise, and the skills of the other staff and the volunteers, the NAHF operations and my life became far better organized and more productive.

As the building was now open, our staff of employees and volunteers continued to grow, and programs were continuously being added, I pressed forward with fundraising efforts in order to support those staff members and the programs. Also, we continued to work with our exhibit company to design impressive interactive, educational exhibits highlighting the aviation contributions of our enshrinees.

On the morning of September 11, 2001, we gathered at our offices for an important exhibit design and construction meeting. We were nearing the end of the design phase, so everyone involved in the process was present. Attending the meeting were Bob Spayd (the NAHF secretary and our volunteer lawyer), Dan Carney (the PPI exhibit company representative), an exhibit consultant from California, and NAHF employees Cee John, Tara Engel, and myself. The meeting began at 8:00 a.m., and a little before 9:00, Bob received a call from his wife frantically telling him that an airliner had just crashed into one of the World Trade Center towers. We quickly turned on the TV in the conference room to see what was happening.

As a history major, I recalled stories of the B-25 accidentally crashing into the Empire State Building while flying in fog in 1945, so I assumed this also had to be a horrible accident. The scene at the World Trade Center was one of confusion and panic. Within minutes of us turning on the TV, a second plane hit the south tower. This was obviously no accident. Soon after, the Pentagon was struck by a third airliner, and we received urgent word that the Air Force Museum and the entire base would be closing, and everyone needed to leave. We, just like the rest of the world, tried to process what was happening as we prepared to evacuate.

I placed a quick call to the Crowne Plaza hotel in downtown Dayton, and they said we could move our meeting to their top-floor conference room. The top floor was not our ideal location in light of what had just happened in New York, but we packed up our exhibit presentations. Then the south tower collapsed. Our hearts sank. We slowed our exit plans and kept watching news coverage, and soon we saw the north tower collapse. By this point, Air Force security forces were making everyone leave the base quickly, so we traveled to the hotel.

Once we settled into the conference room, we attempted to finish our exhibit planning, but the events of the day made it fruitless to continue. Also, we learned that the government was shutting down all civilian air travel. One of our consultants was from California, and Dan Carney was from Oregon. With all flights canceled, the consultant reserved one of the last remaining rental cars in Dayton and offered to drive Dan to California, where Dan could find a way to get home to Oregon.

Dan, however, thought he'd be able to get a flight out of Dayton within a couple of days, so he elected to stay. Commercial flights began after two days, but the backlog of canceled flights meant Dan didn't fly out until a number of days after that. The United States and the airline industry would be forever changed.

Our older daughter, Lori, enrolled at the University of Notre Dame after her high school graduation in 2000. Lori received an AFROTC scholarship, and Karen, Katie, and I enjoyed going to South Bend, Indiana, for Fighting Irish football games as often as we could get tickets. The first game following the 9/11 attacks was played on September 22, and Lori was selected as a member of the honor guard for the flag-raising ceremony.

The display of patriotism was overwhelming, as every person entering the stadium was handed an American flag to hold during the National Anthem. I had chills watching my little girl represent my country and my Air Force as part of a four-person military formation standing alone at the center of Notre Dame Stadium. There is a rumor that I might have shed a tear at that moment, but that can be neither confirmed nor denied. I managed to elude the security people long enough to sneak closer to snap a photo of the honor guard before they sent me back to my seat.

As it turned out, a professional photo of the honor guard, with a background of thousands of American flags in the stands, was turned into a poster and sold at the campus bookstore. Lori gifted us a framed copy with an engraved plate the weekend she graduated, and I had the honor of commissioning her as a second lieutenant in the Air Force. It will always hang in our house.

As our fundraising increased, I hired professionals to produce our annual enshrinement ceremony. Even though our staff, board members, volunteers, and long-time producer Mike White had become pretty savvy at planning and staging the ceremonies, the events became more polished with the professionals' help. In addition to the aviation pioneers we'd always had in attendance—ranging from Neil Armstrong to Chuck Yeager—we also had a celebrity master of ceremonies, such as Harrison Ford, Maureen O'Hara, and Medal of Honor recipient and World War II ace, Joe Foss, who was joined by his friend, baseball Hall of Famer and fellow Marine Corps pilot, Ted Williams. We developed a well-deserved national reputation as the "Oscars Night of Aviation."

In addition to the many speaking and fundraising engagements Tara and I were making, we were offered an extraordinary opportunity that enhanced our efforts. Through the good graces of Russ Hodge of Three Roads Communications, their corporate sponsors, and a generous donation from one of our trustees, Tara and I were asked to participate in a Public Broadcasting Corporation series titled *Legends of Airpower*. This furthered the NAHF's national outreach, and we taped the series for three years during visits to the Three Roads studios in Washington, D.C.

We took advantage of the Washington trips to visit politicians at the Capitol, as well as aviation-oriented businesses. Thanks to our organization's friend, Cliff Robertson—winner of the Oscar for Best Actor in 1968—we delivered lunch speeches to various groups in Washington and, later, in New York City. A fact unknown to most people, Cliff was a licensed pilot who regularly attended the NAHF enshrinement ceremonies and served as our emcee in the past. He even came to Tipp City once to dine with my family at a local restaurant. (Mom was a fan and was thrilled to sit next to him.) Cliff was a strong believer in the NAHF mission and someone I considered a true friend. He was inducted into the NAHF in 2006 for his many contributions to aviation.

The grand opening of the National Aviation Hall of Fame's learning and research center was held on January 8, 2003. Our temporary exhibits had been removed, and the permanent interactive exhibits were in place. All of the trustees and our staff had every reason to be proud of how far we had come in the relatively short span since 1992.

Almost since the beginning of my job as the executive director at the NAHF, I planned to resign once the building was completed and the learning center was opened to the public. Many, including most board members, had often heard me say, "I wanted to build the museum, but I don't want to run the museum." There were several variations on this theme, which I used over the years. "I want to build the Hall of Fame; I don't want to run it," or, "Once the building is done, I'm moving to the parking lot in a Winnebago." I wasn't kidding.

Once the NAHF building was opened and I'd spent a period of time overseeing the operation of the museum, I was even more certain that it was time for me to leave the organization. I wasn't unhappy; I just thought I'd done all I wanted to do, so it was time for me to move on. Bear in mind that, in the Air Force, I'd never been in one place—let alone one position—for longer than three-and-a-half years. The organization was on extremely sound financial and operational footing, so I felt it was time to officially announce my retirement.

After the grand opening, I felt my work was done, and I prepared a letter of resignation to give to the board of trustees. On February 10, 2003, while Don Vanderkarr was in my office prior to a subcommittee meeting with some of the trustees, I handed him my letter and asked him to share it with Clyde Autio, the president of the board. Don was one of the two people who hired me in 1991.

They both were surprised to receive the letter and immediately tried to talk me out of leaving. I told them what I had said to them many times before, "I wanted to build the Hall of Fame but not run a museum."

They said they'd contact other executive committee members and give them copies of the letter, but, in the meantime, "Do not announce anything." I agreed.

Later that day, Clyde returned to the building, closed the door to my office, handed me my letter of resignation, and said, "Don and I don't want you to leave." They felt the NAHF needed me to stay in order to organize the upcoming *Centennial of Flight* celebration—the 100-year celebration of the Wright Brothers' first successful flight in December of 1903. It would be the largest NAHF gathering ever planned, and it was less than six months away.

I reiterated that I enjoyed the fundraising, design, and construction portions of the building project, but I had no interest in the day-to-day operation of a museum. I explained that the noise from some of the exhibits made me nervous, especially the loud, unexpected explosions of the rocket launch demonstrations. (In hindsight, I probably suffered a mild case of Vietnam-related PTSD.) And I did *not* enjoy being the person who had to go down to the exhibit area regularly to chase kids off of the exhibits and maintain order when the volunteers couldn't.

He told me that he and Don understood the demands of my new role but said, "We want you to stay on long enough to train a new executive director." Pause.

"And, oh! Would you be interested in eventually changing your role to being the development director, a position with reduced hours and few administrative responsibilities?" Another pause.

"Oh, yes…one other thing. Would you keep your letter of resignation from the rest of the board? It might cause some unrest regarding the future of funding efforts."

I said I'd consider the development director position but *only* with more flexibility in my hours and IF I could have an office located outside of the actual NAHF facility. My reasoning was that there were no empty offices for use by a new executive director once he or she was trained and actually assumed the role, and I no longer wanted to be around all of the noise and hassles of the learning center. I knew if I stayed in the building, my role would eventually morph into being an assistant to the director.

I also wanted to see if Tara would be interested in starting an independent development company and having the NAHF as our first client. I was good at fundraising and making visits to potential donors, but Tara had the magic touch in writing strategic plans, funding proposals, and grant requests. I also felt I owed it to Cee to see if she wanted to move to the annex office with us, as she was already training another NAHF employee, Kerisa Citro, to back her up at the office.

Clyde agreed, so, with board approval but not board understanding, I began the search for a new executive director, whom the board agreed would temporarily be named *Director of Operations*. Two of our trustees, John Lombard and board secretary Bob Spayd, were tasked by Clyde to locate a small office near the base and the museum that would eventually be a fundraising office for Cee, Tara, and me.

On April 21, 2003, I hired a qualified former Air Force and airline pilot to begin working as the Director of Operations, and he was told he would become the executive director in the near future. In the meantime, with one additional employee, we were out of offices. The newly hired Operations Director took my office, and Tara and I moved to an office building within sight of the base and the museum which the two trustees had leased. Cee was to join us as soon as the Ops Director was up to speed in his new job, if she desired.

Unfortunately, shortly after the man began as our new Operations Director, he suffered a relapse of a serious medical condition. He was forced to seek medical attention and, eventually, was unable to work anymore. It was a sad time for all of us at the NAHF.

Our enshrinement ceremony in July of 2003 celebrated 100 years since the Wright Brothers' first successful flight. We named it the *2003 Pioneers of Flight Homecoming*, and it was a part of the Dayton-wide *Centennial of Flight* celebration. This was the most spectacular enshrinement the NAHF ever staged. It returned twenty-three of our past inductees for our dinner and ceremonies. The Master of Ceremonies was Hollywood superstar and aviation enthusiast Harrison Ford. It was an absolutely magnificent event and the highlight of the *Centennial of Flight* observance.

I think I helped the NAHF during my thirteen years there, but I also benefitted in many ways. I had the opportunity to meet many of the pioneers of aviation whom I had admired both as a kid and as an adult, and many are still friends today. The list of heroes that I met would fill a few pages, so I'll mention just a few: Buzz Aldrin, Gene Cernan, James Lovell, Wally Schirra, John Glenn, Scott Crossfield, Dick & Burt Rutan, Paul Tibbets, Chuck Yeager, and Frank Borman, who was kind enough to write the foreword to my first book, *Naked in Da Nang*.

But three of my most vivid memories of those years involved one man: Neil Armstrong, the first man to walk on the moon. I'd met Mr. Armstrong a few times at events over the years, and I was fortunate to introduce him to my family and have photos taken of him beside Karen, Lori, and her future husband Thomas, Katie, and me. One photo is autographed, which makes it extra special since he rarely gave autographs.

The first memory is from October of 2003. I met Neil (He had now asked me to call him that) in Orlando, Florida, where Tara and I joined his good friend, Harry Combs, to present the NAHF's *Combs Award* at the annual conference of the National Business Aircraft Association (NBAA). I had the honor of introducing Neil at the conference and the opportunity to spend time with Neil and Harry for part of the conference.

How Neil Armstrong and Harry Combs became such close friends is a story few people know, but I was honored when they related it to me one night. Most people know that Neil was a shy person and didn't seek publicity. After his return from the moon in 1969, government officials sought a private place for him to decompress and enjoy time away from the frenzy that followed him continually. Harry offered NASA his huge Colorado ranch as a place where Neil could escape the crowds and avoid the spotlight for a while. Neil and Harry remained fast friends from then until Harry died in 2004.

The second trip I made with Neil was to another NBAA conference, this one in Las Vegas. We stayed at the Paris hotel, and one of the highlights of my life was walking across the casino floor toward the exit and hearing someone yell, "Hey, Mike!" That someone was Neil Armstrong. Wow!

By this time, I had met many celebrities and stars, but this was the absolute best moment ever. Neil Armstrong, a world-famous astronaut and the first man to step on the moon, is calling my name. Did I say, *Wow*?

Neil, Tara, and I walked out of the hotel and got into a taxi to go to the convention center, and I remember thinking, *This taxi driver has no idea he has Neil Armstrong as a passenger*. Along the way, Neil and I got into a discussion about the computing power of his lunar lander versus current technology. I made the statement, "Today's (2003) flip phone probably has the computing power of your lunar lander."

I'll never forget his response. "Today's microwave ovens probably have more computing power than the lander, and the reason the lander's computer kept sending error messages during the landing was the computer couldn't keep up with the inputs." I'll take him at his word. He was there; I wasn't.

The final time I saw Neil was in Denver in 2004 at the memorial service for his friend Harry. Tara traveled with me to the service since she had worked closely with Harry on a large donation. His generosity endowed the *Harry B. Combs Learning & Research Center* and our annual *Combs Award.* Tara and I definitely felt out of place around the many corporate and political leaders in attendance, and no one paid much attention to either of us. *Ignored* would be a fitting description.

A short time into the social hour at the country club, Tara and I heard a mumbling, low at first but growing louder, at the entrance to the room. We'd already retreated to the social no-man's-land at the back, so we couldn't see what was happening, but we quickly learned that Neil Armstrong had arrived at the memorial, and everyone craned to see him.

As Neil took just a few steps into the ballroom, we saw that he looked ill at ease with the crowd. As he scanned the room, he spotted Tara and me and made a beeline to where we were standing. I guess he felt at ease with us because he stayed for a few minutes, talking about our good times together with Harry. After Neil went off to converse with others he knew, Tara and I'd suddenly become socially acceptable, I guess, since other attendees approached us--undoubtedly curious to learn who we were and how we knew Neil Armstrong.

The last time I spoke to Neil was in mid-2007 when I called him (Yes, I actually had his phone number and email address) to ask, "Would you mind if Tara and I write a young-adult-level book on your life?"

He couldn't have been nicer, as always, and responded, "No, I wouldn't have a problem with that. I'd even offer to help you, but I can't. Someone is already working with me on a biography, so I'm probably contractually prohibited from helping to write another book, even a young adult book."

He did offer, "If you have any questions, though, feel free to call or email me."

I thanked him and said good-bye to what turned out to be our last conversation. He was a truly great American, an aviation pioneer, and a wonderful person. Rest in peace, Mr. Armstrong.

Eventually, a current NAHF staff member was selected by the board of trustees to be my replacement. I was honored to be named Executive Director Emeritus, and I left in January 2005 to begin a new and exciting chapter of my life. The National Aviation Hall of Fame was in good shape, and I was proud of what all of us had accomplished in order to transform it into a robust national organization. I was sorry to leave *my* organization, but I'd already begun working on great things to follow.

Presentation of the Purple Heart medal from American Legion post commander, Gene Chaney, twenty years after Mike's injury

Autographed photo of our family with Neil Armstrong and Harry Combs

The grand opening of the National Aviation Hall of Fame at the Air Force Museum at WPAFB

On our front porch in Tipp City while Lori (R) was home from the University of Notre Dame

AFROTC cadet Lori Jackson as a member of the honor guard for Notre Dame's first football game after the attacks of 9/11

Chapter 10

The Book That Inspired the Celebration

While my staff and I worked to fund, expand, and build the NAHF into a truly national organization, I also continued my independent efforts to help veterans, particularly Vietnam veterans, with their needs. I knew many of them suffered still not only from their experiences during the war but also from the unwelcome treatment they received when they returned home. My own memory of returning from Vietnam to be greeted at the San Francisco airport by that bunch of long-haired, profane protesters shouting insults never left me.

My anger toward those protestors only grew over the years to include all anti-military activists and similarly-minded politicians, and that dislike remains to this day. I know I am not alone in this feeling. A retired general who attended an event where I was the guest speaker shared with me that his wife and infant daughter met him at the airport upon his return home as a captain from Vietnam, and all three were spat on by proponents of "peace and love." Forty years later, I saw his eyes mist over and his fists clench at the recollection.

My experiences with others who were mistreated upon our return from Vietnam—and for years after—have shown me that the veterans who suffered the most were the ones who quickly returned to civilian life when they set foot on American soil. Too many of these loyal men and women found they were ridiculed, spoken down to, turned down for jobs due to anti-military bias, and, in general, unsupported by the society for which they risked their lives

Some returnees were "separated at the port." They went from a dangerous war zone, fighting for their survival or that of their comrades in arms, on one day to being civilians with an uncertain future less than a week later. Many returnees had no job lined up and felt they had no support from their previous employer: the United States government. Add to that their realization that sections of the American public had been conditioned by the media and politicians to distrust the military, and it's no wonder some veterans had trouble adjusting upon their return home. I remember hearing about a restaurant that had a sign on its door that read "No GIs or Dogs Allowed." Why were we treated so differently than returning WWII and Korean War veterans had been?

The group of Vietnam veterans who fared best was those who stayed in the military or entered jobs with a similarly rigid structure, such as law enforcement, firefighting, and security. I attribute this to the fact that they were around others who shared a respect for authority, observed a clear chain of command, were pro-military, and had a strong sense of order and mission. That's just my opinion, but, to me, it has been borne out by years of experience. I do know that remaining in the Air Force after my return from Vietnam made my readjustment easier than for those who dove directly into civilian life.

When I returned from Vietnam in 1972, I began writing a book relating my experiences leading up to and going to Vietnam. My vision was for it to be a *true* representation of the Vietnam veteran—not the trash that was being put out by Hollywood or the liberal media. It would be the true story of one guy from a small town in Ohio who went off to war and returned, relatively in one piece, both physically (other than the whole getting hit with a rocket thing) and mentally. I knew my story was not much different from the experiences of thousands of others who answered the call of duty and came home to normal lives, but it was far different from what we saw on television and on film.

Also, my book would take a different tone, as I wanted to relay some of the funny things that happened to me along the way. Definitely a different approach from what was on the market. However, life being what it is, I never got too far in my writing.

During my time as the NAHF executive director, I occasionally told Tara and others some of the funny (or dumb) things that happened to me in the military. One day, after hearing a particularly engaging series of laugh-out-loud tales of flying low and slow over Southeast Asia, Tara said, "You need to write a book."

"I started a book once."

"When?"

"About thirty years ago."

"That's great! I want to read it. Can you share it with me? How much do you have done?"

"Oh, probably five or six pages."

She assured me that mine was a story that needed to be told and asked, "Could I help you write it?"

After quickly calculating that, at the rate of six pages every thirty years, I'd probably never finish, I eagerly said, "Yes!"

So began Tara's and my joint effort to write a true and sometimes-humorous book about how I ended up in Vietnam and then returned home. I didn't want to be accused of doing this "on company time," so we'd meet for breakfast at the local Bob Evans restaurant before starting our work day (which explains why the Bob Evans restaurant is included in the book's acknowledgments). We also worked on it many weekends. About two years later, *Naked in Da Nang* was published in June of 2004.

As our writing progressed, Tara and I talked extensively about the hostile reception Vietnam veterans received upon their return. Each of us, at different times, wondered to each other if our book couldn't be the catalyst for an actual national event to finally and formally welcome home those veterans—or if it was too late. As we continued, the idea of a national welcome home never left our minds. *Wouldn't that be wonderful?* But how could you ever direct something that great from the little town of Tipp City, Ohio?

To ensure the book was as factual as possible, Tara and I interviewed more than two dozen Vietnam veterans and other people with whom I'd served in the military. Something that deeply affected Tara was how the Vietnam veteran and I would have a discussion for the book, and, at the conclusion of the interview, we'd shake hands and say "Welcome home" to each other.

After Tara heard that exchange following a number of interviews, she finally asked, "Why do you always end the interview with that phrase?"

I explained, "No one ever welcomed us home, so it's something Vietnam veterans started doing among themselves."

By this point, I was on the verge of crying, and Tara had passed that point completely. I'd never taken the time to examine what the lack of a proper welcome meant to other veterans or to me, and it broke my heart to think about it in depth. It meant so much to hear "Welcome home"—even from another Vietnam veteran. It was all really very sad. I think that it was at that point that Tara and I both stopped talking about a national welcome-home celebration as a concept and started trying to figure out a way to make it a reality.

Tara and I sent the final manuscript of our book to the publisher and then focused on the preparations for July's enshrinement ceremonies. I'd turned in my resignation at the NAHF in early 2003, but I was staying until my replacement was hired and trained.

In May of 2004, I commissioned Lori as a second lieutenant during graduation weekend at the University of Notre Dame. She reported to her first assignment at Seymour Johnson AFB in North Carolina in July during the same week the first copies of *Naked in Da Nang* arrived. It was exciting for me to finally hold the book. If nothing else, it would be a good memoir to share with my wife and daughters. I had no expectation that it would go on to touch a national audience.

Our first sales of the book were to NAHF board members during our July 2004 hanger party at the Moraine airport on the Thursday before enshrinement weekend. A month later, we took copies to the annual Experimental Aircraft Association fly-in at Oshkosh, Wisconsin, and took turns selling them at a table outside the NAHF exhibit when we weren't working on the NAHF clock.

Most of our early sales were to fellow veterans who were probably intrigued by the title *Naked in Da Nang: A Forward Air Controller in Vietnam*. The first indication that we might be onto something that truly touched Vietnam veterans was when a big, burly ex-Army veteran came back to us two days after buying a book and asked, "Who gave you permission to write my story?"

After I recovered from being almost scared to death and hiding behind Tara, he smiled and explained, "Your story is much like mine—my childhood and my military experience, except for the branch of service and occupation." Then he bought four more books, "Three for each of my adult children and one for my ex-wife so they might finally understand what I've been through."

We've learned over the years since publication that Vietnam veterans have trouble talking about their experiences. I know I had very few serious conversations about it, even with my family, until we started writing the book. Funny stories I could do, but I could never talk about "the serious stuff." Our book gave veterans a vehicle to tell their story—even if it was from an Air Force pilot's perspective—by simply handing the book to someone and leaving it with them to read. I think the fact that it had almost no vulgarity and no braggadocio helped make it relatable to most veterans. Most of us weren't Rambo. Most of us were clean-cut, normal people who'd been thrust into an abnormal environment.

A second veteran returned to buy extra copies for his family since he thought it would be a book in which they could envision his experiences, which he'd been unable to relate to them over the years. Both Tara and I were taken aback by the reactions of the veterans since we'd gone out of our way to avoid a "blood and guts" war story and, instead, concentrated on the thoughts of a typical Vietnam veteran in the years leading up to and going to Vietnam. As a matter of fact, the only criticism of the book in any of the online reviews was that it didn't have enough action. These reviewers didn't understand the reason why we wrote the book and what it was meant to portray. Actual Vietnam veterans did understand, and this understanding led to more than we ever could have imagined.

Katie and Mike during Dad's Weekend at Ohio University

Some of the staff at the Bob Evans restaurant near WPAFB before Mike gave a Veterans Day speech near the base

Lori's commissioning ceremony at Notre Dame. Mike was honored to administer her oath of office.

Second Lieutenant Lori Jackson and Lieutenant Colonel Mike Jackson, USAF (Retired)

Chapter 11

The First National Operation Welcome Home

In November of 2004—a few weeks prior to my leaving the NAHF—some of the staff, including Tara and I, traveled to one of our annual events: the Thunderbird Hanger Party and the annual air show at Nellis AFB in Las Vegas. Some of us normally went to help set up for the party, meet potential supporters, and man a display table at the air show to promote the NAHF and its activities. What began as just part of our NAHF job became so much more.

On one of the days during the air show, we were lucky enough to meet with representatives from the base, the Fremont Street Experience, the Las Vegas Convention & Visitors Bureau, and a few casino owners. They had heard about our book and our idea for a national welcome home celebration from Bob Avery and some others with Aviation Nation, the organizer of the annual air show, and they wanted more information.

Shortly after we returned to Ohio, the people from Aviation Nation and Las Vegas—a truly patriotic city—notified Tara and me that they would be honored to host a national Operation Welcome Home (OWH) event in 2005. "Oh, by the way—you, Mike, would need to serve as the national chair, and Tara would be co-chair.

"And Las Vegas would host the event and provide logistics and other support, Aviation Nation would handle Las Vegas and air show coordination, and Nellis AFB would provide the organization and site for the air show." How could Tara and I say *No*? The answer: we couldn't. This was our dream opportunity.

The remainder of the national organization, publicity, and speaking engagements would be arranged by Tara and me, with little salary or travel budget. So, I stepped into a new title: *National Chairman of Operation Welcome Home*.

The events were scheduled around Veterans Day in November. That meant there was a *tremendous* amount of work to accomplish in the eleven months prior. Aviation Nation was, thankfully, well-entrenched in the air show business in Las Vegas and at Nellis AFB, but the planning of events for veterans started from scratch. Tara and I needed to create some type of organization to provide national publicity as well as plan for veteran events, all while figuring out how to survive financially.

When I left the NAHF in January of 2005, Tara stayed and planned to simultaneously serve as the co-chair of Operation Welcome Home. However, it soon became obvious that her trying to coordinate our national OWH events with Aviation Nation and with me by phone and email when her work day was over wasn't going to work.

Therefore, Tara resigned from the NAHF in mid-January of 2005 in order to devote full time to her role as co-chair of OWH.

We were thrilled to be provided with a space for Operation Welcome Home at the Tipp City Chamber of Commerce's office suite until we could find a place of our own. Shortly after, we were able to rent a small office upstairs in the Monroe Township building, the same building the Chamber of Commerce occupied. This became the new national headquarters of Operation Welcome Home. *National headquarters? Whoa! What have we gotten into?*

To keep us afloat financially while we worked on OWH publicity and funding, we started what turned out to be a successful fundraising and consulting business, Integrity Marketing & Communications, LLC, which worked primarily with aviation and military museums across the eastern United States. In addition, we started a non-profit 501(c)(3) organization, The American Veterans Institute, to continue our support of veterans in need and also provide free consultation to veterans who needed help with Veterans Administration paperwork and problems.

As our book sales increased and our independent fundraising and consulting efforts grew, we began to travel across the country to publicize the upcoming OWH celebration in Las Vegas. An example of one month's travel can be seen in our May 2005 schedule. The first week: Washington, D.C., for G. Gordon Liddy's national radio program, a speech at the Washington Press Club luncheon, a presentation at the Air & Space Museum, and a presentation at the Pentagon. The second week: presentations at the Atlanta Vietnam Veterans Businessmen luncheon and at the National Aircraft Finance Association in Florida. The third week: the Mississippi Airports Association. And, in the last week of May: the Selfridge ANG Base air show in Michigan. I sometimes felt as if I was living out of a suitcase.

Traveling and speaking to groups was necessary to get the word out to our veterans, with only a few months before November's celebration in Las Vegas. An interesting thing came out of our radio interview on G. Gordon Liddy's nationally broadcast radio program. Mr. Liddy was so impressed with what we were doing for Vietnam veterans and our knowledge and commitment to helping *all* veterans that while he was on the air in the weeks following our appearance, he called us at our office on a few occasions to ask about various programs for veterans and how to file for the benefits. If we didn't know the answers immediately, we'd call Denver Combs, our friend and fellow veterans advocate, to get the answers.

It would be safe to say that neither Tara nor I profited from the success of our book, *Naked in Da Nang,* as most of the money from sales went to promote OWH activities. And, thanks to the generosity of all of the people and organizations in Las Vegas and at Nellis AFB, all of the events of the four-day OWH celebration were to be free to participants. No one thought Vietnam veterans should pay for their own long-overdue welcome home celebration.

One potential hiccup to the otherwise smooth planning arose a few days before we left Ohio when we received a disturbing email from one of the individuals whose organization had been helping to organize the parade. "We're afraid the parade will be too long, so we're going to limit the number of people registering to march."

What?! Both Tara and I were totally aghast, to use a polite word! We, alongside many, many, many organizations, had been working for nearly a year to make this the welcome home parade that Vietnam veterans deserved and would *finally* receive after almost forty years. Then to have them arrive and be told the parade was full and they wouldn't be allowed to participate? No way could we allow that! It would be the ultimate insult to veterans who anticipated their long overdue welcome and traveled miles at their expense to be told, "Sorry, we're full. Watch from the sidewalk." Unacceptable in *so many* ways! We were incensed!

We prepared for battle. If Aviation Nation and others couldn't change those minds, we'd see what we could do to prevent this injustice. Tara and I immediately pleaded our case to representatives of the offending organization—with no effect. They stood their ground. We, however, secretly stood *our* ground and decided we were *not* going to let this happen. We came up with a plan.

"We'll stand by the registration table, smile, and tell veterans, 'You don't need to register. Just join in the parade,'" said Tara.

"Sounds good," I agreed. "Why would it be necessary to register to simply march in a parade, anyway?"

Next, we drafted and printed dozens of press releases, which Tara would carry with her when we left for the parade, to be released to the press in attendance. "I'm more than willing to be arrested for refusing to allow the registration of the veterans," I assured her. "It'd be GREAT publicity for our cause." I almost hoped they *would* arrest me!

A few days before Veterans Day in November of 2005, Tara and I, along with Karen, Lori and her husband Thomas, Katie, and a contingent of Tipp City and Ohio friends, joined thousands of Vietnam veterans from across the nation (and some from other countries) who headed for Las Vegas for *Operation Welcome Home*—a four-day, free celebration honoring our nation's Vietnam veterans. Tara and I had traveled the country promoting the celebration, and we were cautiously optimistic we had churned up enough interest through news releases, media interviews, and personal appeals to many groups across the country to guarantee a good turnout.

To label it "a good turnout" wouldn't come close. It was a HUGE turnout of veterans and their families from all parts of the country. They were greeted in a way that only Las Vegas could greet you—with signs, billboards, banners, radio and TV messages, and even skywriters. The Freemont Street Experience enthusiastically turned itself over to the OWH celebration and designated various stages for entertainment as well as designing a special blocks-long overhead light show for their Vista Vision, the world's largest digital screen, to honor Vietnam veterans.

The four-day celebration began on Thursday, November 10th, with a number of bands performing on Freemont Street Experience stages. A large replica Vietnam Memorial Wall was opened with a ceremony by Native Americans who performed a purification ceremony honoring the veterans who had died in service to their nation. I played only a small part in the planning of this ceremony, and until then, I was not aware that Native Americans were among the staunchest supporters of the military during the Vietnam War. These fine warriors later performed a powwow at a nearby stadium to honor *all* living and deceased Vietnam veterans.

The only pay-to-attend event was a Thursday evening dinner at the Thunderbirds' hanger on Nellis AFB. This was the annual Hanger Party, and while it wasn't open to the public, it provided us with an opportunity to thank all of the sponsors of the celebration and honor the Thunderbirds and Nellis AFB. The Thunderbirds were to be the special performers at the airshow on Saturday and Sunday of the OWH weekend.

Among the attendees at the dinner was General Robin Olds, one of the heroes of the Vietnam war. Tara and I had known him for years, and, to us, he represented the true "warrior spirit" of those who served in Vietnam. We had previously spent an evening with General Olds listening to his amazing stories of flying in World War II and in Vietnam. Entertainment was provided by the Johnny Crawford Orchestra. Johnny was a Mouseketeer in the 1950s, a recording star in the 1960s, and starred as Mark McCain in the TV series, *The Rifleman*.

Friday, November 11th, Veterans Day, started out early for Tara and me. We had to be on Freemont Street at 4:45 in the morning for a fast cycle of morning news show interviews for a number of TV stations from across the nation. We remained seated in the same chairs as people would run in after one interview, change our microphones and TV cameras to a different station's and tell us who we'd be talking to, and then we'd do another quick, live interview to be seen in some place in some different time zone. When we finished the series of TV interviews, it was off to the start of the Operation Welcome Home parade route and waiting for the start of the parade. Tara assured me that copies of our press release were in her pocket.

We showed up early at the location for forming the parade and found that a registration table had, in fact, been set up, and a number of veterans had already registered. We immediately stood a few feet in front of the table and began telling approaching veterans, "Hey, you don't need to register. Please spread the word to others for us."

We never actually told anyone *why* the table was even there or *why* to skip registering, but our plan worked. Few people registered, and we didn't get any blowback from those working the table. After all, what could they do? The parade formed as planned, and I wasn't arrested. No one besides Tara and I knew about our victory, but we felt *so* victorious.

That was the only potentially negative issue in the glorious four-day celebration. Tara and I were all smiles as the parade got off to a tremendous start under a beautiful Las Vegas sky.

And what a parade it was! Just fantastic! There were more than one hundred veteran and veteran-oriented units, buses, military vehicles, and bands joining the thousands of Vietnam veterans. Leading the parade was Las Vegas Mayor Oscar Goodman, with Tara and me following closely behind as the chair and co-chair of the celebration.

This is the only parade I've ever heard of that was larger at the end than at the beginning, and this says something about the need to have this celebration. Since I was at the front of the parade, I could see many people lining the parade route who wore various types of military clothing, whether it was a *Vietnam Veteran* hat, a fatigue shirt, or a jacket with military markings. I didn't do anything as the parade started, but I quickly realized that these veterans had probably been traumatized by the negative reception upon their return from Vietnam, and they didn't want to be insulted again if the parade went badly and disruptors started yelling or interfering with the parade in any way.

I broke out of the parade and told one of the first veterans I saw, "This is your welcome home parade. Come join us." Then I actually pulled him into the parade. He was so overcome that he started crying and thanked me as he joined "his" parade. That started a wave of people already walking in the parade urging those on the sides of the street to join them for "their parade."

There were people in bleachers, others waving from multi-storied parking garages, and spectators ten deep along the entire parade route. There were skywriters flying overhead proclaiming, "We (heart) our veterans" and "Welcome Home."

By the time the parade ended two-and-a-half hours later, it was much longer than when it began, and the participants were both emotional and happy. There were many tears shed in joy and handshakes and "Welcome home!" greetings from within the parade and from those on the sides of the street. I'm biased, but I don't think any parade could have been more successful as a tribute to—and properly welcoming home for the first time—our Vietnam veterans.

Mayor Goodman, Tara, and I stood at the end of the parade route, trying to shake as many hands as we could. I didn't know Mayor Goodman, and I still don't, but he impressed me as someone who truly appreciated our veterans. There are few politicians today who would stand in the heat for two-and-a-half hours to thank people who weren't eligible to vote for them.

Still reveling in the glow of the parade, Tara and I went to one of the stages that Freemont Street had reserved for us and gave a talk to the audience telling why and how Operation Welcome Home was envisioned. We thanked everyone for coming to the celebration and outlined the events for the remainder of the celebration. Then, the bands performed for the rest of the day and into the night. The fabulous Freemont Street Experience light show was shown continuously throughout the weekend.

Saturday and Sunday were air show days for veterans and for the public at Nellis AFB. Nellis and Aviation Nation had organized an outstanding air show that featured dozens of aerial acts, with a focus on Vietnam-era planes in addition to jet acts and, of course, the Thunderbirds. There were many ground exhibits, and many of those were of Vietnam-era planes and vehicles.

A highlight of the air show for me and for all veterans was the presence of Ann-Margret, a Hollywood star and a performer in Bob Hope's USO tours to Vietnam in the 1960s. Tara and I had the chance to talk with her, and she was delighted to receive a copy of *Naked in Da Nang*. She then offered to have a photo taken with me while she held a copy. She later took the air show stage and gave a heartfelt talk to "my boys," as she frequently refers to Vietnam veterans. I know all of us in attendance came away from her show feeling she was truly one of us, as you could feel her sincerity as she spoke to the crowd.

The next day, Sunday, November 13th, was the final day of the Nellis air show and the final day of Operation Welcome Home events. It had been a wonderful, if exhausting, four days of around-the-clock events and celebrations. I saw so much happiness and heard so many expressions of appreciation from the attendees to all of those who had helped to make this celebration become a long-overdue reality. There were so many people whom I tried to thank before I left and after we returned to Ohio, but I'm certain some deserving people were missed, so let me thank you now on behalf of a generation of truly deserving veterans who appreciated your help and your words of thanks.

On Monday, the entire Ohio delegation of my friends, my family, Tara, Denver Combs, and I headed home for some well-deserved rest and recuperation. Our Operation Welcome Home celebration was a success, over and done… or was it? We soon realized that our supposed one-time Welcome Home event would ripple across the nation for years to come.

Thank you to the city of Las Vegas, to the casinos who supported OWH, to the Fremont Street Experience, to the countless military and veterans organizations, and, of course, to Aviation Nation and to Nellis AFB for making this a truly memorable experience for all who attended.

The official insignia of Operation Welcome Home

The start of the Operation Welcome Home parade in Las Vegas

The Native American purification ceremony of the Vietnam Wall replica in Las Vegas

Mike and Tara Dixon Engel at the start of the OWH parade in Las Vegas

Ray Santana, participating in the OWH parade in Las Vegas; he would go on to co-chair the OWH celebration in California a year later.

The Las Vegas chapter of the Military Order of the Purple Heart was one of more than one-hundred marching units in the OWH parade.

One of the many crowded venues along the parade route

Ann-Margret was a guest of honor at the Nellis AFB airshow and posed for a photo with her copy of Naked in Da Nang before performing for the crowd.

Chapter 12

The Ripples Widen

When Gary Chapman found out Las Vegas was going to host an Operation Welcome Home celebration for Vietnam veterans, he knew what he had to do. He emailed his friend Ray Santana.

Ray, I just read a news article. They're going to have a Welcome Home parade for Vietnam veterans in Las Vegas on Veterans Day. I want you to walk with me in that parade.

Little did he know what a difference that one email would make—not just for Ray and for himself but for thousands of Vietnam veterans that year and in the years to come.

Ever since Ray returned from a year serving with the Marine Corps in Vietnam in 1968-1969, he refused to talk about his war experiences. When Linda met him in 1997, it was clear to her that he suffered from PTSD. As she learned more about his year in the conflict, she wasn't surprised.

Ray served as an infantry machine gunner. His duties ranged from manning a flame thrower to anti-tank and demolition assignments. On Ray's worst day, he was caught in the sight of a Viet Cong's rifle. The sniper fired three times, once grazing Ray's head, but the worst result was the spray of sand into Ray's face by a bullet that struck the ground in front of him. Despite being bloodied and blinded, Ray wasn't med-evacuated that day. Casualties had been high, and his unit needed him, so he stayed with them and shared night watch.

Ray couldn't see the enemy, but he could hear when they approached. He noticed the frogs and crickets silenced, and then he heard splashes. "Wake up! Wake up!" he whispered to the others in time for them to fire at the approaching Viet Cong. When the Marines' bullets hit the explosives which these enemies were carrying, the threat was eliminated in a big way. By waking the others, Ray had undoubtedly saved many lives. He was evacuated two months later due to a spinal injury, and, unfortunately, his medical records fell from the back of the helicopter as it took off. The nineteen-year-old's life was changed forever.

Since no written record of the injuries he received in battle survived, Ray received no medical or compensation benefits when he was discharged. It wasn't until 2006 that the first benefits arrived, following a diagnosis of PTSD, and later, he was declared 100% disabled following a diagnosis of prostate cancer that was declared service-related. He has had eleven back and five neck surgeries in an attempt to heal his physical wounds. Mental wounds are more difficult to heal.

Gary served as a door gunner on an Army helicopter when he was nineteen. His unit was known as the "Warlords," and it earned a reputation for the "snatch"—a rapid evacuation of men or equipment. Gary was also wounded in battle, and he received two Purple Hearts. Upon his return home, he became active in local veteran groups, eventually serving on the Los Angeles County Veterans Services Board.

Chapman's hometown of Palmdale, California, and Santana's hometown of Lancaster, just to the north, are veteran-friendly locations, as are most areas around military facilities. The cities are located in northern Los Angeles County in the Antelope Valley of the western Mojave Desert. Edwards AFB lies about thirty miles northeast of Lancaster, and the extensive Navy facilities in San Diego reinforce the citizens of southern California's positive attitude toward military veterans. When the mayors of Lancaster and Palmdale found out that Ray had agreed to accompany Gary to Las Vegas to walk in the parade, both mayors named the men ambassadors from their towns.

So, Gary, Ray, and their spouses, Ingrid and Linda, traveled to Las Vegas and participated in the various OWH activities Veterans Day weekend in 2005, and when they returned to southern California, they were different people. Having finally seen such a massive display of kindness and appreciation for all Vietnam veterans left them feeling pride for the efforts and the sacrifices made by so many brave individuals. Both Gary and Ray shared their stories about their experiences in Las Vegas with all who would listen. The welcome home might have come thirty-eight years later than it should have for them, but its impact couldn't be denied. They felt the healing begin for their wounds on the inside.

Soon after the Santanas returned from Las Vegas, Linda contacted Tara and me. She wrote, "Your celebration in Las Vegas changed Ray's life. Now he wants to do the same for others."

Tara sent Linda information about the OWH movement and included a photo of the parade. Unbeknownst to Tara, the photo was of Ray! The die was cast. California needed its own Operation Welcome Home!

Ray, Linda, Gary, and Ingrid got to work to make that happen. They quickly sought help from many sources, and the citizens of the Antelope Valley answered their call. Tara and I remained in contact, offering advice and suggestions throughout the planning.

Support came from all directions. The Santanas and the Chapmans began contacting veteran groups and media outlets throughout the Valley as they set their sights on an Operation Welcome Home celebration on Veterans Day that year, 2006. Gary and Ray had numerous contacts in veteran groups, and both Ingrid and Linda had media experience and contacts. Their friend Ida Ketchum, who had written the newspaper article Gary saw, was retired from a NASA unit located on Edwards AFB, and she had access to base personnel.

Ida had learned about the book *Naked in Da Nang* from her sister Frieda, who lived in Troy, Ohio, near Tipp City, and Ida gave a copy to Dennis Anderson, the editor of the *Antelope Valley Press*. Anderson was himself an Army veteran of the war in Iraq, and his son was a Marine assigned to the 4th Infantry Division in Iraq and then Afghanistan from January to June 2006. Anderson publicized the California plan, and the ball was rolling.

The base commander at Edwards AFB promoted it, as did the mayors of both Lancaster and Palmdale. The father of Palmdale's mayor was killed while fighting in World War II, so his son did all he could to help with logistics.

In the midst of the planning, a complete stranger called the Santana home. "Do you want a Huey? I know how to get one," she explained to the planners, "and I want to take a ride in the kind of helicopter my father flew in." The caller's father had died in such a helicopter in Vietnam, and this was her contribution to honor his memory.

Linda's mother was part Cherokee and part Algonquin, and Linda had listened to stories about code talkers when she was a child. She got in touch with a group of code talkers—some from WWII and some from Vietnam—who resided with the Navajo Nation in Window Rock, Arizona. There were also many Vietnam veterans among the Navajo.

Terry Norris, a Navy and Marine veteran and owner of Precision Labs, a Palmdale calibration laboratory serving both military and civilian industries, is rightly proud of his own Navajo heritage. "My father was the first Navajo to earn a Ph.D.," he told us. He read about the local committee's idea to bring the Navajo code talkers and veterans to town, and "I decided it was something that I wanted to help."

Through calls to the *Antelope Valley Press* and, then, to the reporter, he tracked down Linda's contact information and gave her a call. He offered to pay for the Arizona contingent's travel and hotel expenses if they would make the trip for the parade. Prior to the Navajo group's arrival, Norris found out there was only a small budget for their food, so he spent more than thirty hours setting up a lavish military banquet in their honor, which included a fallen soldier table.

Tara and I flew to California on November 8 to help with the final preparations. I spoke to broadcast reporters in interviews to publicize the event.

The morning of November 11, 2006, began unusually cold. As Linda, Ray, Ingrid, Gary, Tara, and I arrived at the staging point on Lancaster Boulevard, we saw no activity. *What's going on?* we wondered. By the time the parade program was printed, ninety-three parade entries had pre-registered, and nearly one hundred Vietnam veterans registered to march. Plus, members from all branches of the military would follow them. Where were they?

At the starting time, we got our answer. Cars were parked along all of the cross streets in order for the parade participants to stay warm. Before long, the groups and individuals were assembled. Time to begin.

First, the Huey from Wings and Rotors Air Museum in Murrieta, California, flew over the route and continued to do so, start to finish, finish to start, throughout the parade. Then Ray and Gary led the Vietnam veterans who had registered. Their ranks soared as Ray pulled reticent bystanders into their midst, just as had happened in Las Vegas. He said after the parade, "I did everything I could to get Vietnam veterans to leave the sidewalk and join the parade." Wives helped to push others into the stream.

One hundred ten units followed. These were active-duty units from Edwards AFB, high school JROTC units, middle and high school bands, Gold Star Parents and Blue Star Mothers, Patriot Guard riders, Girl Scouts, Boy Scouts, and numerous civic and veteran groups. The Navajo code talkers and Vietnam veterans marched or rode behind a banner which identified their group. Tara and I shared the spotlight, following a sign identifying us as the National Chairmen of Operation Welcome Home.

Perhaps most moving of all were the individuals who had lost a loved one in Vietnam. They wore t-shirts with WALKING FOR THOSE WHO CAN'T emblazoned on the front, and they each carried a photo of the departed.

At the end of the parade route, the organizers greeted the parade participants and their families and friends. Ray, Gary, Linda, and Ingrid were told numerous times, "My husband is talking to the other veterans." The wives of the code talkers said, "Our husbands are smiling." These veterans finally felt the comradery back home that had once been theirs as they fought for survival in the jungles of Southeast Asia.

Linda Santana says, “It showed how hungry Vietnam veterans were for the recognition.”

“The timing was right,” says Anderson. “Recognition was afforded to the full field of service. Using Vietnam as the hub, all veterans were included in this recognition, including Korea and Iraq.”

At the banquet for the Navajo code talkers and veterans that evening, Norris was profoundly moved when he witnessed the appreciation shown by his guests from Arizona. He was presented with a poster they’d made, on which they’d all signed their names. It continues to hang in his office today, along with the traditional arrows and artifacts he’d also been presented.

That night, Linda escorted the Navajo chief back to his hotel. As they were saying their final goodbyes, the chief placed his hands on Linda’s shoulders, looked her straight in the eyes, and said, “Little one, you have a good sleep.”

Without question, the four who had worked tirelessly for months to make their dream a reality slept well that night. However, the stone we’d thrown into the pond in Las Vegas sent out more and more ripples. The work for veterans was nowhere near finished. Like Tara and I, the Chapmans and Santanas were on an emotional journey that changed their lives… and the lives of countless others.

That’s what the Operation Welcome Home movement was all about. As Dennis Anderson put it, “Vietnam vets felt validated and appreciated.”

Terry Norris was inspired. “People need to become involved,” he told himself. “It was the changing point for my life.” From that day on, he has sought many opportunities to become more and more involved in his community, including serving in local civilian and military organizations and in positions on numerous boards. “I never knew it needed to be done.” He continues these efforts to this day and credits the Operation Welcome Home celebration as the start.

Two years later, a two-act play entitled “A Piece of My Heart” was staged by a local theatre group. It is the true tale of six women—five nurses who were serving in Vietnam and a country-western singer who came to entertain the troops. Those who attended found the performance powerfully moving, and it became a "must-see" experience in the Valley. Everyone seemed to be talking about it.

On the same weekend, a traveling Vietnam Memorial Wall visited. Some members of Point Man Antelope Valley (PMAV) were present. Point Man International Ministries (PMIM) is a nonprofit organization with the purpose of "addressing and meeting the spiritual, physical, and emotional needs of veterans and their families." These local members agreed, "Why don't we build our own replica?"

Why not? With those two moving experiences during that one weekend, the timing was right again.

Efforts began to amass the $100,000 that would be needed for building The Mobile Vietnam Veterans Memorial for the Antelope Valley, or "AV Wall" for short. When the $80,000 mark was reached, contributions stalled. Linda suggested, "How about we involve children by having them collect Pennies for Soldiers?" She called her son-in-law, who contacted the PTA at his daughter's school, and West Wind Elementary School was on board.

Eventually, twenty-seven schools throughout the Antelope Valley joined the efforts during the school year of 2008-2009. Each school was encouraged to "Adopt a Soldier" from the Wall, one of the seventy-plus young men from Antelope Valley. For example, West Wind selected Army CPL Terry Joe Bailey, who died in April of 1971 at age eighteen. Students researched Bailey, learning all they could in order to put a face on just one of the 58,000 names engraved on the Wall. Lesson plans were written to highlight the deeds of veterans, and enthusiasm and support grew.

The Pennies for Soldiers grass-roots effort in those 27 schools raised $20,000 in pennies collected in jars placed in classrooms, resulting in the goal of $100,000 being reached. The Wall was built—a large replica of the memorial on the mall in Washington, D.C.—and it continues to be displayed every year in Palmdale on Veterans Day. The name of every school which participated is included on a panel that lists all donors and accompanies the AV Wall in its travels. Point Man Antelope

Valley is responsible for the maintenance, storage, and displaying of the Memorial.

Things seemed to slow down after that until one day in 2018 when a school administrator approached Linda, asking, "What can we do now? Those students who worked so hard ten years ago are now near or past graduation. We need something to inspire our new, younger students."

Linda's suggestion was an essay contest about the importance of remembering our veterans. The winners read their essays at the AV Wall on the next Veterans Day.

And so, the momentum continued. In 2019, volunteers gathered to create a special community project to be included as part of the public display. Participants created a 12-foot by 3-foot Vietnam service ribbon to be placed near their replica wall, a "Welcome Home" sign for all veterans who pass by and who visit The AV Wall.

And the ripples expanded. The celebrations we set in motion in 2005 had spread to California and beyond by 2006 and would continue across the country for years to come. The philosophy *If we could help just one more…* was alive and well in Ohio, Las Vegas, and Southern California, and it continued to grow across the country, wherever military veterans and those who love and respect them are found.

Posing with one of the flower girls before the Antelope Valley, California OWH parade

Gary Chapman (left) and Ray Santana (middle) shake hands with one of the Navajo Code Talkers at the OWH celebration in California.

Navajo ladies with their handmade Navajo Nation Warriors Code Talkers and Vietnam Veterans banner at the start of the parade

Chapter 13

Indiana and Beyond

Sometimes when you least expect it, things just happen to fall into place and work out perfectly. Such was the case after Tara and I returned from California, and our for-profit company, Integrity Marketing and Communications, became our focus. Since we both had a successful history in running a museum space and in raising the operational dollars we needed to do so, we specialized in fundraising for museums—many times, military museums. That was the case when we received a call from the World War II Victory Museum in Auburn, Indiana.

Dean Kruse was the president of the Dean V. Kruse Foundation, which operated the museum. He had earlier acquired the contents of a military museum in Belgium, which contained one of the largest collections in the world of primarily German WWII military machines. When the Belgian museum ran into financial difficulties, Kruse purchased and moved 140 military vehicles, 50 motorcycles, and hundreds of military uniforms to new construction he'd built in Auburn. Why Auburn? And how does one amass the fortune needed to purchase such an inventory—and then ship it to Indiana? Interesting story.

First of all, one needs to know a little about Auburn, a town of about 13,000 residents in northeast Indiana. The town has a rich automotive history. In 1904, a horseless carriage factory was built there. That business evolved into the Auburn Automobile Company, whose prize model was the Auburn, a fast, luxurious, good-looking, and expensive vehicle. Other models were the Cord and the Duesenberg. Auburn's historic reputation for upscale automobiles made it the perfect setting for Kruse's operations to thrive.

How did Dean Kruse rise from his humble beginnings on a dairy farm near Auburn to amass sufficient funds to purchase the entire contents of the museum in Belgium? That took some time and effort.

In 1952, Dean Kruse, Sr., became an auctioneer, and his son Dean liked that business much more than farming. He began to help his father in his business at age 12, and he earned his own auctioneering certification at age 16. At age 18, Kruse started to work full-time in his father's business, which flourished by selling everything from artwork to islands to airports to entire towns.

In 1971, Kruse and his father created an annual Kruse Collector Car Auction, which they ran together. Kruse bought out his father and brother in the mid-1980s, and the auto auction's success grew. Always the brilliant entrepreneur, Kruse gave away 30,000 tickets for the auction annually to local residents. None of those folks had any intention—and, of course, the wealth needed—to purchase the sometimes exotic, always expensive automobiles brought in from far and wide to be sold to the highest bidders. The serious bidders feared there had to be someone else in the vast audience who could outbid them, so the bids went up and up—far higher than if there had been just a small audience comprised of only the serious bidders. Pretty smart move.

The interest in Kruse's annual auction grew even more when it was televised nationally. Therefore, it came as no surprise to most people when Kruse sold the entire operation to eBay in 1999. The reported (but unconfirmed) price... $165 million. More than enough to finance new ventures, which included his vision for a National Military History Center, of which the WWII Victory Museum was the first step.

In November of 2006, Bob Krafft, the executive director of the World War II Victory Museum, contacted Tara and me. He explained, "We want to discuss the possibility of your raising funds for our museum's operations."

We eventually met Dean Kruse, and, in discussions with both Dean and Bob, we learned Dean's goals for the center. "This project isn't about celebrating war… just the opposite," he began. "Really, it's about honoring the rich legacy of the people and the events that have preserved our freedom." Tara and I liked the sound of that. This was definitely something we could support, so we agreed to help raise operating funds for this vast, impressive site. We got to work, both at home and during numerous trips to Auburn.

One of the benefits of going to the Kruse Museum was getting to climb on and inside the tanks, trucks, and various other war machines early in the morning before the doors opened to the public. To make navigating around the one-hundred-plus displays easier, the museum had acquired an early version of a Segway, a two-wheeled, gyro-stabilized, self-balancing personal transporter that I'd hop on whenever I could "borrow" it and race around the museum… "Are you sure that's a good idea?" Tara questioned, but I ignored her, of course. Happily, I never ran into anything expensive.

In the process of working with the Kruse staff, Tara and I often spoke of various Operation Welcome Home regional and local events we were coordinating at the same time. The more we spoke to Bob Krafft and his staff, the more we all began to see a perfect connection between the goals of the museum's educational and patriotic programs and our goal of honoring America's veterans. The museum's staff was enthusiastic about the positive effect that widespread publicity surrounding a large national veterans' event could have to further the goals of their organization. Holding a five-day national Operation Welcome Home event from July 2 through July 6, 2008, in the midwestern town of Auburn, Indiana, would be a win-win for both of us.

While we concentrated on fundraising for the museum, we also helped the staff begin to organize our third large-scale, national OWH event. The Kruse staff was just as enthusiastic about the event as Tara and I were, and they all put in extra hours gathering support, organizing participants, and lining up donors. Since the site was a military museum, a number of local and state veterans' organizations were knowledgeable about its operation, and some were already involved in its activities, so veteran support was easy to gain.

As in California, individuals and organizations jumped on board to help. Local military vehicle clubs and reenactors were eager to participate in the celebration, and other organizations offered to contribute food and soft drinks, which would be plentiful and free. And, unlike many of our other celebrations, we had the luxury of time to get everything organized before the event.

The Operation Welcome Home celebration in Indiana was different from those in Las Vegas and California in several ways. The most significant was the expansion to a five-day format, of course, but that wasn't the only difference. There was a wide range of activities filling those days.

Veterans were welcomed into both the WWII Victory Museum and the Kruse Automotive & Carriage Museum free of charge during the entire five days of the celebration. The Kruse staff and their families had also agreed to incorporate a fundraiser for a Fort Wayne charity, The Shepherd's House, whose mission was to fund apartments for homeless veterans.

Barbara Cox, who operated The Shepherd's House with her husband Lonnie, told Tara and me, "Facilities for veterans are few and far between. We believe the homeless veteran issue is a moral and patriotic crisis for our generation… and the solution is to give these homeless heroes the love, attention, and support they have earned."

Her desire was to raise enough money to build an additional suite onto her facility. Tara and I agreed that we had a unique opportunity to give back to the region that opened its arms to us. I told Tara, "Barbara and her all-volunteer staff are changing lives and making a difference every single day. If we can help them with their mission, then we are reinforcing our mission as well. It's a perfect fit."

It was easy to plan events to fill the five days. We incorporated an essay contest for students in grades seven through twelve, held prior to the event. Also, there were numerous presentations held inside the museum to inform veterans of benefits for which they might be qualified but unaware. And, of course, Tara and I conducted book signings for *Naked in Da Nang*.

On the first evening, July 2, the Freedom Ride motorcycle processional, sponsored by a nearby Harley-Davidson dealership, escorted the Traveling Vietnam Veterans Memorial Wall across northeastern Indiana to the grounds of the museum. Bob Krafft led the procession, riding on a custom red-white-and-blue motorcycle that belonged to the museum. A variety of military vehicles followed The Wall. Live music and plenty of food for the riders and visitors followed into the night.

The Wall set-up began early the next morning, continuing as a convoy of Vietnam gun trucks made their way across northeastern Indiana. A variety of elaborate exhibits opened on the grounds of the museum, and early afternoon saw a Blessing of the Wall ceremony.

The only dark spot (literally!) on our celebration came in the evening on July 3, when a severe thunderstorm passed near Auburn, knocking out power. We had to put the final touches on the upcoming events in the dark, working with flashlights. However, the next day's sunrise brought perfect weather, so life was good. All systems, go!

Friday, July 4, was the busiest day by far, with events from 9:00 a.m. until dusk. Live music filled the air as the American Huey 369 helicopter arrived, a Vietnam-era vintage Huey that flew attendees up and over the Vietnam Wall replica for the next three days. (Bob, his father, and his son all got a ride.)

Mid-afternoon saw the presentation of our Welcome Home Essay Contest finalists, followed by a moving Memorial Service held to honor veterans, providing an opportunity to reflect on their service and sacrifices. The keynote speaker was Indiana State Senator Allen Paul, a decorated Vietnam veteran himself, who served with the Army's 1st Air Cav division. At dusk, the city of Fort Wayne's Independence Day Fireworks lit up the sky.

The aforementioned Huey rides and exhibits continued on the 5th and 6th. Other events included a concert provided by the Air Force Band of Flight from Wright-Patterson AFB in Ohio in the evening on July 5. Kiwanis Club fireworks closed the day in the Kruse Auction Park at 10:30. On the final day, July 6, there was an amazing demonstration of the power of tanks when they crushed cars.

Throughout all five days, donations were collected for The Shepherd's House. Cans were set inside the museum and on a table outside. Periodically, either Tara or I would empty the cans and keep track of how we were doing. Whenever the pace of donations seemed to slow, we sent one of Bob Krafft's young daughters and a friend, carrying a cute puppy and a donation can, into the crowd to solicit more. Who could possibly resist them? Quite an effective move!

Thanks to the hard work of the staff, the generosity of the attendees, and the irresistible cuteness of Bob Krafft's daughters and their young friends who roamed the museum grounds seeking donations, the goal for housing was reached. We arranged a ceremony at the conclusion of the event to present a check to The Shepherd's House, which continues its efforts to this day.

As Bob Krafft now reflects on the day, he says, "It was an honor to have that celebration in Auburn—in our little part of the world." Kraft and his staff and volunteers are some of the most patriotic people I've ever met. It was an honor for us, too.

So, The Indiana OWH event was yet another successful—and free—celebration to honor both Vietnam veterans and veterans of all eras. When the weekend drew to a close, it was back to work at our regular jobs for the Kruse staff and for Tara and me.

The Kruse Museum fundraising was ongoing. Tara and I were able to procure funding from foundations, private individuals, and groups, as well as funding from the state of Indiana. We were also successful in expanding their educational programs. Tara and I wrote three veteran-oriented children's books with grants from several sources to place the books in schools in Ohio and Indiana, along with lesson plans and other suggestions we developed to accompany the books.

In the years following the Auburn, Indiana, OWH, Tara and I concentrated on our American Veterans Institute program, which offered aid to veterans, while continuing to help others organize Welcome Home events across the country. Tara and I were able to attend some of the continuing OWH celebrations, and some we weren't.

A 2011 article in the *Dayton Daily News* prior to an OWH event in Dayton, Ohio, in November of that year called Operation Welcome Home "the celebration that won't end." We hoped they were right.

Banner announcing the Auburn, Indiana OWH

Hungry veterans and supporters awaiting food in Auburn

Huey helicopter flying above the replica Vietnam Memorial Wall in Auburn, Indiana

Check presentation to the Shepherd's House.

Chapter 14

Recognition for Veterans Grows

In the years 2005 through 2008, our lives were dominated by the major Operation Welcome Home celebrations in Las Vegas, California, and Indiana, but Tara and I were busy on other fronts as well.

Even before the Las Vegas celebration, our speaking engagements with local venues, as well as coverage within local and national media, led to a few local and regional OWH events. Following the Las Vegas celebration, Tara and I were inundated with inquiries from people across the country asking us for advice on how to organize their own OWH events.

Tara's and my for-profit company, Integrity Marketing & Communications, had been our source of income since we left the National Aviation Hall of Fame almost two years prior. It was doing well, despite our main focus being the OWH celebrations. We were proficient by this point in raising funds for veterans' organizations and museums.

A key to our success was the approach we used with our client organizations, which was different from what I used when I started as the one-person staff in my early days at the NAHF. Back then, being inexperienced at any type of fundraising, I hired a couple of professional fundraising companies. My experience with these organizations was this: they came in and charged us for general consulting advice, literature, and services. Then I did most of the work, and the fundraisers left before they'd raised enough to cover even their consulting fees.

For my second round of interviews with professional fundraisers, I asked them at the start, "Why can't I pay you based on how much you raise?"

Their reaction was akin to my asking them to perform a violent criminal act. "No way. That would be impossible."

That did it. I was done hiring consultants and set out to raise funds on my own—and I became quite good at it. Later, when Tara joined the NAHF, she brought another component to our fundraising efforts with her skills in writing strategic plans and creating funding proposals.

When we formed Integrity Marketing in 2005, we took the lessons we'd learned at the NAHF but added the component of charging only a set (small) percentage of what we raised for our client. Beyond that, the only thing we charged for was writing a strategic plan if they needed one. Bottom line: if we didn't raise any money, the client organization would pay nothing. In the years we ran Integrity Marketing, we never failed to meet our client's goals, so both our clients and we did quite well.

Our second area of focus was strengthening our veterans' service organization, the American Veterans Institute (AVI), a non-profit IRS 501(c)(3). Our primary goal was to provide free help to veterans, assisting them in gaining the benefits they had been unable to get on their own.

The second goal of AVI was to promote future Operation Welcome Home events, large or small, across the nation. We had begun this while we prepared for the Las Vegas event, as the word of OWH spread through our presentations and articles in 2005.

The two goals became intertwined. While on speaking tours about upcoming OWHs, as well as during the actual celebrations, we met so many veterans whose requests for benefits had been turned down by the Veterans Administration that we felt we had to do something to help. We did refer many to the Veterans Service Officers in the major veteran organizations, such as the VFW and the American Legion, but sometimes the individuals we spoke to just needed extra, personalized help. We, the AVI, charged nothing for our services, and we became experts in helping veterans with their appeals after they had been turned down by the VA.

We helped veterans in a variety of ways. Local veterans tended to come to me since they were usually WWII veterans who knew both my father and me. I assumed they were more comfortable talking to and working through their issues with me since I was a friend and also a "local boy" versus a Veterans Service Officer they didn't know. One example of our efforts—and our successes—was a WWII veteran who was finally awarded a Purple Heart in 2008. He had been trying to get it for more than sixty years. Incidents like that just hardened our resolve to help anyone we could.

Another area where we focused was the Improved Pension Program. At the time, this was a little-known VA program which provided funds and/or support to veterans in need under various circumstances. We were fortunate to have our veteran benefits expert, Denver Combs, volunteer his expertise whenever we ran into an area in which we lacked knowledge. We created a checklist for applying for this program that, in our opinion, "magically" appeared on the VA website in a format nearly identical to ours. That was perfectly OK with us; anything that assists veterans in their attempts to get the benefits they deserve is a good thing.

[Editorial comment: the American Veterans Institute website currently found on the internet uses our original home page and photographs, but it is no longer *our* site, and we have not used or updated the site in more than ten years. It is not overseen, nor is it endorsed by Tara or by me, and we have formally requested it be taken down.]

Our third area of concern, on a local scale, was establishing a veterans' memorial park in Tipp City. Ron Re, a retired Tipp City police sergeant and the city's only two-time Citizen of the Year, envisioned the park being placed on a city-owned vacant lot on a Main Street corner, but he'd run into opposition regarding the site he'd selected. To most people in town, it seemed the perfect location. A memorial there, surrounded by a small, carefully landscaped park, would be seen daily by everyone driving through town. However, for whatever reason, Ron ran into trouble trying to convince the Tipp City Park Board and, in turn, the Tipp City Council members to approve this location. Both entities, which included only one military veteran, advocated for a site in one of the existing city parks. In Ron's and our opinion, a memorial structure placed in one of those parks would be out of sight to the majority of citizens for most of the year, and it had the potential for being misused or vandalized due to the remote location.

Ron asked Tara and me, as local veteran advocates, to help him convince the Park Board and Council to allow the memorial to be constructed on Main Street. This might seem to be an easy problem to solve. It was not. We worked with Ron on a publicity campaign that resulted in 2,000 signatures (from a town of about 10,000) and letters of support from twelve civic and social organizations. It took us more than a year to get permission to construct the memorial on the desired site, but the will of the people couldn't be ignored. As a measure of public support, fundraising for the memorial reached nearly twice the original goal, allowing for additional benches, a partial payment for a gazebo, an irrigation system, and additional landscaping.

The Tipp City Veterans Memorial Park in its Main Street location proved to be a valuable community asset upon its dedication in 2007, and it continues as such. It turned out beautifully, and it has become a site for weddings and photos. More importantly, it is the location where students plan and gather annually for a Veterans Day program and where the city's Memorial Day parade begins. It's located on the perfect site.

While we confronted these three areas, we coordinated requests from localities across the United States for information on how to stage their own Operation Welcome Home celebrations. We were thrilled that the publicity leading up to the Las Vegas celebration and after the event itself had created so much pro-Vietnam-veteran "buzz" and support. We were happy to answer all of the phone queries, and initially, we mailed packages with tips on how to gather local support and funding, how to publicize the event, and how to organize a parade and other activities. Eventually, we created a website that had all of the information anyone would need to plan and conduct their own community's Operation Welcome Home celebration.

We attended OWH events when we could and received positive feedback from the organizers of other events. We have been asked many times how many OWH events came about through our help, and the simple answer is… we don't know. Some calls, emails, or web contacts might have led to actual OWH events, and some might not have. Our best guess is that there were thirty or more events held across the country while we were helping organize OWH events. But, honestly, we're not totally certain when OWH began. We know that our first official, national-scale OWH was the one in Las Vegas in November of 2005, but we had begun to publicize Vietnam welcoming events at the end of 2004, and our efforts were featured in magazines and newspaper articles by February of 2005.

Whether there's a direct connection or not, we do know that Troy, Ohio, a town neighboring Tipp City, named Vietnam Veterans as the Grand Marshals of their annual Strawberry Festival in June of 2005, five months prior to OWH Las Vegas. And, in the fall of 2006, Tipp City named "Saluting Our Veterans" as the theme of their Mum Festival (small towns in Ohio like festivals), and I rode in a convertible to accept a check for the town's veteran's memorial at the reviewing stand. Was there a connection between OWH and either of these events? Possibly, but they weren't named Operation Welcome Home. Do they count as "official" OWH events? Who knows, and who cares, as long as veterans are being honored?

We did our best to make it as easy as possible for any community to hold an OWH event, from answering questions to suggesting potential sponsors and fundraising ideas when asked. We were advised before the onset of the Las Vegas event to trademark the term Operation Welcome Home and our logo (which had been graciously designed for us by our friend Heidi Burton), but we said no. We wanted everything we developed to be available for other celebrations with no restrictions. Our only hope was that events would be free to the veterans.

As time passed and we received feedback from veterans about the success of the celebrations, we received one piece of advice that we used from 2007 until we ended our participation in the Operation Welcome Home celebrations. The advice: add more unrecognized veterans as part of the celebrations. We realized that

Korean War veterans had not been honored upon their return after their service in that war. They may not have been met with the insults and abuse the Vietnam veterans encountered, but they'd been largely forgotten. In subsequent celebrations, we recognized and honored Korean veterans along with Vietnam veterans.

Later, we were made aware that the veterans of the Cold War had not been honored for their service. I certainly could relate to this. As a B-52 pilot, I had been one of the crewmembers who were part of the "nuclear triad," the three-pronged nuclear deterrent made up of missiles, submarines, and strategic aircraft on 24-hour alert throughout the Cold War. Military veterans who served across the world as part of the "trip wire" to the Soviet threat also deserved recognition, no matter where they served. So, we added the Cold Warriors of the nuclear triad as well as all those who had served our country during that period.

We remained determined to recognize Vietnam veterans but to never forget any veteran who served and did not receive the recognition they were due. Of course, the Vietnam Service Medal ribbon remained in our logo, and it remains at the center of our hearts.

Tara and Mike starting another day at their auxiliary office, table 44 at Bob Evans in Tipp City

The Veterans Memorial Park on Main Street in Tipp City, Ohio

Lori and Mike pointing to their memorial bricks near those of relatives at the Tipp City Veterans Memorial

Chapter 15

Final Thoughts

I've been asked many times why Tara and I elected to undertake such a huge project as Operation Welcome Home alone instead of combining efforts with any of the major veteran organizations. The simple answer is that we did initially contact a couple of the major veteran organizations; however, as in dealing with all large, national organizations, waiting for an answer while the discussions crawled through their various committees and sub-directorates took too long. Bear in mind that we had only eleven months between our leaving the NAHF and the November 2005 Las Vegas celebration and about nine months in 2006 after we were contacted about an Antelope Valley celebration.

I have compared waiting for a large organization's decision to turning the Titanic or turning a B-52; it takes a long time and a big area to make the turn. On the other hand, we, as a two-person partnership, could turn on a dime, so we were able to make decisions quickly. We didn't wait for a committee's decision. We just didn't have the time to wait, so we did it ourselves—with a whole lot of local help, of course. Without local groups and volunteers in Las Vegas, California, and Indiana, those celebrations wouldn't have been possible. In all three cases, Tara and I were in Ohio. We answered questions and offered suggestions, but local veteran and civic groups—as well as individuals—did the work and the fundraising.

By 2012, our efforts were starting to be overtaken by Welcome Home events organized by larger pro-veteran organizations to honor returning veterans from Iraq, Afghanistan, and the other "hot spots" where the United States has been actively or covertly involved in the decades since 9/11. Since we had purposely not trademarked any aspects of our OWH celebrations and had, in fact, posted planning tips and staging formats for all to see, we were pleased to see any and all Welcome Home events honoring veterans, whether we had a role in them or not.

Tara and I feel we succeeded in our goal of honoring Vietnam veterans initially and all veterans by the end of our participation in celebrations in 2012. Through an estimated thirty or more Operation Welcome Home events ranging from local parades to our national events, Vietnam veterans were given the opportunity to hear the country finally say a collective "Welcome Home" and "Thank you for your service." For two people working out of a one-room office in small-town Tipp City, Ohio, I think we did pretty well.

However, the need for all of our veterans to hear appreciation from our nation's citizens continues. Starting with the end of the draft in 1973, the percentage of American citizens who volunteer for service in the military has steadily fallen and now represents only about one-to-two percent of eligible citizens. This tiny percentage is what keeps us free and forms the tip of the spear to face all military and often humanitarian burdens for our nation and the free world, as well.

Military members are underpaid, and enlisted personnel sometimes have to rely on food pantries to feed their families. This is a disgrace! The least we as a nation can do is support the active-duty military and their families by pressuring the well-paid, overcompensated politicians to increase military salaries so we don't have families on food stamps while their service members could be placed in harm's way anywhere around the world.

The military is not an easy life for service members or their families, and they need the country's continual support. NEVER allow the policies of our politicians to undermine the continual service and sacrifice of our military and their families.

Every member of the military has their own motivation for joining the Armed Forces, but there are common threads that seem to bind the members together. I hope the story of my experiences conveys just how human our guys and gals in uniform really are.

To ALL veterans, thank you for your service and "Welcome home!"

Receiving an honorary doctorate degree during the June 2012 graduation at Ohio University

Zieg Family

Patel Family

Our whole family at Christmas in 2021. Karen and Mike are seated in the middle, with the Patels (Katie, Akash, Kedar, & Aryanna) and Ziegs (Lori, Thomas, Allison, Andrew, and Caitlin) surrounding them.

Glossary

AFMPC – Air Force Military Personnel Center - Sometimes shortened to MPC. Headquartered at Randolph AFB, Texas, it is the personnel center where most officer and enlisted jobs and assignments are made. The name was changed to Air Force Personnel Center in 1995 to include civilian personnel actions.

AFSC – Air Force Specialty Code - An alphanumeric code used by the Air Force to identify specific jobs. Officers have 4-digit codes, and enlisted have 5-digit codes. AFSCs sometimes have letters or numbers before or after to further identify a duty position.

Alert – A readiness level in the military used to denote a unit or aircraft crew pre-positioned and ready to engage in offensive or defensive action. Nuclear alert had crews manning missile silos, ground and sea units, and aircraft within minutes of activation.

ATC – Air Training Command - One of the major commands of the Air Force. The training arm of the USAF, headquartered at Randolph AFB, Texas. It was re-designated Air Education and Training Command following a merger with Air University in 1993.

B-52 – Often called "the BUFF" (big ugly fat fellow) - An eight-engine heavy bomber that can be used in both strategic and conventional roles. The H-model had a crew of six in 1972.

BOQ – Bachelor Officer Quarters - On-base living quarters for bachelors or unaccompanied married officers

FAC – Forward Air Controller - Pilots flying to protect and support friendly troops by directing air strikes against enemy forces. FACs were used extensively in Vietnam.

GI – Slang term for military personnel derived from the term Government Issue

MEPS – Military Entrance Processing Station - Joint service locations across the U.S. where potential military recruits for all branches of service are tested and given their initial physical examinations

NAHF – National Aviation Hall of Fame - Established in 1962 with the mission of "honoring America's Air and Space Pioneers."

O-2A – A twin-engine (one in front, one in back) light aircraft used by forward air controllers during much of the Vietnam War

OTS – Officer Training School - The 12-week course for earning an officer's commission. Previously located at Lackland AFB, Texas, and now located at Maxwell AFB, Alabama

PAS – Professor of Aerospace Studies - The commander of an Air Force Reserve Officer Training detachment at a civilian university

PCS – Permanent Change of Station – A permanent move to a new duty station usually lasting two-to-four years but could be longer or shorter

PIT – Pilot Instructor Training - A 12-week course to train rated Air Force pilots to be Undergraduate Pilot Training (UPT) instructor pilots

PQ – Pilot Qualification - A program to train rated pilots as certified pilots in a particular aircraft

SAC – Strategic Air Command - One of the major commands of the Air Force. It included the nuclear flight and missile arms of the USAF. SAC was disestablished in 1992, and its personnel and equipment were redistributed to the other major commands. SAC's Offutt AFB headquarters was transferred to United States Strategic Command (USSTRATCOM).

Standardization/Evaluation or Stan/Eval – The unit at bases throughout major air command levels which is responsible for ensuring units are properly trained and operating according to directives. It administers ground and flight certifications to rated pilots.

TAC – Tactical Air Command - One of the Air Force's Major Commands through 1992 when it was incorporated into Air Combat Command

TASS – Tactical Air Support Squadron – In Vietnam, one TASS was assigned to each of the four corps (geographical subdivisions) of the Republic of South Vietnam and to Thailand to provide direct air support of combat operations.

TDY – Temporary duty - A temporary duty assignment other than the location of a service member's permanent duty location

Tweet – A nickname for the T-37 side-by-side jet trainer aircraft. Also called the "Tweety Bird" or "the "6,000-pound dog whistle to turn fuel to noise" because of its loud, piercing engine noise

UPT – Undergraduate Pilot Training - Formerly a 53-week intensive training program to train USAF pilots and award their pilot wings and rating. Training aircraft flown during this program included the propeller-driven T-41, the twin-jet T-37, and the supersonic T-38.

Acknowledgments

Karen Jackson, my wonderful wife of 53 years and co-author of this book. Thanks for the many sacrifices you made supporting my military career and the family burdens you shouldered while I was off "doing my Air Force thing."

For all of your love, support, and sacrifices over the years:
Lori Jackson Zieg and Thomas Zieg
Katie Jackson Patel and Akash Patel

Tara Dixon Engel: Thanks for many years of fun, hard work, support, and success. To misquote someone: *You did everything I did, dancing backward and in high heels.*

Col. Dale Ullrich, USAF (Ret)

Lt. Col. Bill Blackmon, USAF (Ret)

Lt. Col. Ken Hamlin, USAF (Ret)

Lt. Col. Loren Stendahl, USAF (Ret)

Lt. Col. Bob Watson, USAF (Ret)

Operation Welcome Home, Las Vegas:
Aviation Nation
Ann-Margret Olsson
The Freemont Street Experience

Joe Schillaci, former, Pres & CEO The Freemont Street Experience
Nellis Air Force Base, NV
Las Vegas Convention & Visitors Authority

Operation Welcome Home, California:
Linda & Ray Santana
Ingrid & Gary Chapman
Terry Norris
Dennis Anderson

Operation Welcome Home, Indiana:
Dean Kruse
Bob Krafft
The staff of the Kruse WWII Victory Museum

All of the groups and local organizations in all of the places where there was an Operation Welcome Home or celebration honoring our nation's military

Denver Combs for many years of advice on veteran benefits, filing compensation claims, and for being a good friend

Theresa Lusczek – enshrinement photos from the NAHF
Mike Ullery – enshrinement photos from the NAHF
Heather Schmid for Mandy Paige Photography – Zieg family photograph
Sara Stewart – Patel family photograph

Col. Don Vanderkarr, USAF (Ret), Chairman Emeritus, NAHF
Amy Marino Spowart, President & CEO, NAHF

And, especially, my brothers and sisters in arms who served in Vietnam and Southeast Asia and all those, past and present, serving in our military across the world

About the Authors

Mike Jackson

Mike grew up in Tipp City, Ohio, and graduated from Ohio University with a B.S.Ed. in 1968. He entered the United States Air Force in 1968, earned his pilot wings, and flew fighter, observation, bomber, and training aircraft in addition to serving in a number of command positions, retiring as a lieutenant colonel and command pilot. He flew 210 combat missions in Vietnam, earning the Distinguished Flying Cross, the Purple Heart, nine Air Medals, and numerous other decorations.

Following his 23-year Air Force career, Mike served thirteen years as the Executive Director of the National Aviation Hall of Fame at Wright-Patterson AFB, Ohio. He was named Executive Director Emeritus upon his departure to become the National Chairman of Operation Welcome Home in 2005.

An active alumna of Ohio University, Mike served on their Aviation Advisory Board for the School of Aviation at Ohio

University for eight years, served as the chairman of the Promise Lives Campaign for the Patton College of Education, and served on the Board of Directors of the Ohio University Alumni Association.

In 2004, Mike published the critically acclaimed Vietnam memoir *Naked In Da Nang*. He has since co-authored two additional books: *The Wright Brothers: First in Flight* and *Neil Armstrong: One Giant Leap for Mankind* and has edited four additional aviation-themed books. He appeared as narrator on three seasons of the PBS/Discovery/Military Channel series *Legends of Airpower* and on the Smithsonian Channel's *Unbelievable Flying Objects.*

Mike earned an MBA from Webster University in 1981 and was selected as their Alumnus of the Year in 2005. He was awarded the Medal of Merit from Ohio University in 2007 and was honored with an Honorary Doctorate of Public Service degree in 2012.

Karen Jackson

Karen graduated from Brooklyn High School near Cleveland before attending Ohio University, where she met Mike. After earning a B.S.Ed. degree in elementary education, Karen and Mike married and moved to Texas the very next day. They have been married for 53 years.

As Karen moved around the country with Mike, she taught in six school districts in three states, including South Carolina, Texas, and Ohio. She earned her M.Ed. in Curriculum and Instruction from Texas Tech University, where she worked as a research assistant and taught education methods courses for a brief period before moving again.

Following her move to Tipp City in 1989, Karen taught sixth grade in Tipp City for 22 years, specializing in English Language Arts. In 2005, she earned National Board Certification in Early Adolescent Language Arts, and she retired from classroom teaching in 2011.

In the fall of 2012, Karen became a University Supervisor for the College of Education at Wright State University, a position she held for eight years. As such, she supervised teacher candidates who were completing various phases of field experience and student teaching in public schools.

Karen has served as a valued volunteer for several local organizations, including the Tipp City Public Library Board of Trustees for ten years, the Tipp City Tree Board for five years, the Christmas in the Village committee for eight years, and more than twenty years with The Clothesline, a non-profit resale shop.

Made in the USA
Monee, IL
03 November 2026

67381637R00128